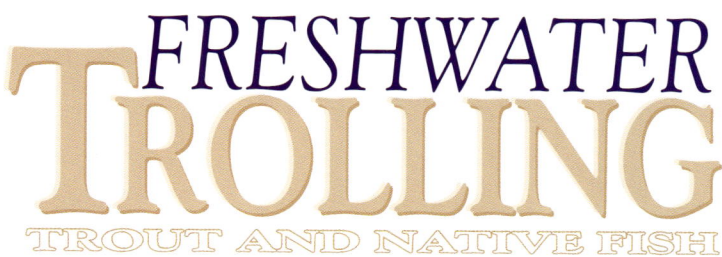

BILL PRESSLOR

Cover Caption: Planer board trolling on Lake Eildon. PHOTO: BILL CLASSON

Published and distributed by
Australian Fishing Network
48 Centre Way
Croydon South, Vic. 3136
Telephone (03) 9761 4044   Facsimile (03) 9761 4055
Email: sales@afn.com.au
Website: www.afn.com.au

This book is copyright. Except for the purpose of fair reviewing, no part of this publication may be reproduced or transmitted in any form or by any means, electronic or mechanical, including photocopying, recording, or any information storage and retrieval system, without permission in writing from the publisher. Infringers of copyright render themselves liable to prosecution.

ISBN 18651 30680

© Bill Presslor, Australian Fishing Network

All rights reserved
First published 2004

FRESHWATER TROLLING
TROUT AND NATIVE FISH

BILL PRESSLOR

Acknowledgments

The production of any book is never as simple a task as it first appears, at least for the uninitiated such as myself. Attempting to translate years of practical experience into words and images is not quite as easy as we might think. I certainly owe a debt of gratitude to Helen Classon and Jim Harmon from AFN for editing, direction and help in making this book a reality. My thanks also to Chris Larkin for her moral support and Robyn Mills for her help with word processing and her extraordinary skill at interpreting my handwriting.

My thanks also to Steve Williamson, Rick Huckstepp, Jim Harmon, Bill Classon and Helen Classon for their photographic contributions, and Dave Silva from Lowrance Australia for his technical expertise with sonar and GPS.

Contents

Acknowledgments iv

Introduction vii

Chapter 1 Boats for Freshwater Trolling 1

Chapter 2 Locating Fish 9

Chapter 3 Lures for Trolling 13

Chapter 4 Surface Trolling 33

Chapter 5 Planer Board Trolling 45

Chapter 6 Getting Down, Aids to Deep Trolling 55

Chapter 7 Trolling Attractors 67

Chapter 8 Downrigging Techniques 77

Introduction

Despite the many technological advances in recent years tolling is a technique that is still under valued as a means of catching fish. Many anglers troll for at least some of the fish they catch and yet others have the attitude that trolling is not a productive technique. I would suggest that those anglers who dismiss trolling as a viable technique are either ill informed or very inexperienced.

Modern trolling techniques allow anglers to target fish that are out of reach by any other technique. Trolling can provide any angler with the scope to be as sporting or productive as virtually any method. Light line and tackle can be employed if you choose and this can increase the enjoyment of landing big fish.

The growing popularity of cast and retrieve methods in freshwater, especially for our native species, has seen many anglers move away from trolling. Nevertheless, most successful anglers who I have talked with still rely on trolling when other methods are unsuccessful.

One of the most undervalued trolling techniques available today to anglers has to be downrigging. Downriggers have been around now for close to 40 years, yet I am still amazed at the number of anglers who have not embraced this technique. I've often heard the argument that downriggers are too expensive, yet the cost of a downrigger is little more than a good quality rod and reel. Properly looked after a downrigger should last for longer than most rods and reels. No trolling technique can provide an easier, more reliable means of targeting fish almost anywhere in the water column. Another misconception seems to be that downriggers are only for trout or salmon. Not so, they are just as effective for targeting natives and virtually any other fish that will take a lure.

Trolling is an effective, challenging method of catching fish in freshwater. If your interests include learning more about fish species and how to locate and catch them by trolling then I hope this book may be of help to you. In nearly fifty years of fishing I have had the good fortune to be able to travel and fish in a lot of different locations. One of the real pleasures of this fishing experience (in addition to catching fish of course) has been meeting a lot of anglers who have been willing to share their experience, knowledge and skills. I hope that I can do my part by sharing what I've learned.

CHAPTER 1
Boats for Freshwater Trolling

Buying or setting up a new boat creates a lot of excitement and anticipation for most of us. Along with all the excitement comes a new set of opportunities, challenges, and decisions. The dilemma of what to choose to accommodate our own particular style or types of fishing can be a daunting task, especially if you are new to the game, but with a little planning and analysis, these tasks can be a lot of fun.

My definition of a good fishing boat for freshwater fishing is one that can provide an uncluttered fishing platform, and I have a few important criteria to judge how comfortable and easy to fish from a boat is. Good storage for fishing and associated gear (fewer things to trip over), some protection from the elements, and a reasonable ability to handle rough water are essential requirements for any good craft. Often we lose track of the fact that whatever type of hull design we choose is most likely going to be a compromise. No one single type or style of hull design is going to give us exactly what we want. The same also applies to the interior design and layout of all boats.

Look for a hull design and interior design that will give you the greatest number of features that you require for your style of fishing and the water conditions you usually fish in. For trolling and downrigging some of the major issues to look at

Smaller boats such as this 4.3m V-nose punt can be fitted out with a vast array of gear and still allow two anglers room to fish comfortably.

include storage, rod holders, some protection from the elements and rough water capability.

BOAT SET UP

When you buy or set up a boat for freshwater fishing think carefully about how you want to use the boat. If you are going to do a lot of trolling and downrigging, storage space will be vital. If you think about the sort of gear you take on a normal fishing trip things start to add up in a hurry. Life jackets, ropes, anchors, cameras, tackle bags, rain gear, fire extinguishers and nets all eat up an unexpected amount of space. It's more than a bit frustrating to get your new boat home, complete with all those numerous hatches, to find that most of them are just too small to be of any practical use!

CHOOSING AN APPROPRIATE HULL DESIGN

The style or type of hull you choose to fish from is generally governed by several factors including your experience, where you fish and the species you target, predominant weather conditions and often most importantly your budget. Trolling can be accomplished in almost any style of craft, but you can make your experience more productive and enjoyable by choosing wisely when you decide on the type of hull you'll use. If your passion is deepwater trolling in some of our larger impoundments, where

Small aluminium boats can be a good choice for anglers who fish protected waters. Boats in this category are easily transportable and very affordable.

weather conditions can mean you will be facing strong winds and rough water much of the time, a deep v-hull may be a smart choice. Conversely, if you spend most of your time trolling in shallow water protected areas, a small aluminium boat in a modified v-hull or a v-nose punt may be the ticket.

Consideration also needs to be given to internal configurations such as seating and consoles. A walk around centre console is a pleasure to fish from if you are anchored and bait fishing, but will it give you the greatest number of options for your style of

Aluminium V-nose punts, such as the well fitted out example pictured here, have become one of the most popular freshwater vessels in Australia.

fishing? Will this type of set-up give you the ability to provide yourself with some protection from the elements? What about access into the boat from shore or from a jetty? Can you launch this style of craft on your own or is it too big to be handled by one person? All of these issues need to be considered when deciding what to select as a new hull.

The advent of the v-nose punt in its many variations has had an enormous impact on freshwater fishing in Australia. It is a stable fishing platform and is ideally suited for impoundment and sheltered water fishing. The development and subsequent popularity of this style of craft has enabled a lot more people to enjoy fishing in our freshwater rivers and impoundments.

This style of craft however does not always afford the greatest protection from the weather. Though its low freeboard generally makes this style of boat less prone to being blown about by strong wind, the trade off is that its seating configuration and fit out usually make it difficult to fit a bimini top, windscreens or other protection from the elements. Boats with low freeboard also don't allow you to lean into the gunnel when fighting a fish or bracing yourself against its gunnel in wind or rough water conditions.

It may seem from these criticisms as though I'm not a fan of this style of boat, but in reality I am. It is a great stable fishing platform, especially if you spend the majority of your fishing time casting and retrieving lures. Just keep in mind that regardless of what type of boat you choose, the height and width of the gunnels on your boat can make a big difference to how comfortable, and easy to fish from your boat really is.

ENGINE OPTIONS

Once you have settled on a hull that is suitable for your style of fishing, think carefully about how you will power the craft. World wide environmental legislation, in an attempt to limit the pollution of our waterways, has had a huge impact on marine engines. All new marine outboards have to be able to comply with these regulations and this has resulted in greater fuel efficiency, ease of operation and reduced noise levels.

Whether you opt for one of the new generation two-stroke engines or choose a four-stroke, think carefully about how you will use the engine. If you are serious about your trolling, how slow you can troll with the engine is probably going to be more important than wide-open throttle speed. The latest generation of four-stroke outboards can provide the angler with the ability to troll at idle speeds for extended periods with little or no consequences to the overall performance of the engine. Though generally more expensive in any given horsepower range, four-stroke engines are a pleasure to use.

TECHNIQUE SPECIFICS
(TOOLS FOR TROLLING)

Any discussion about trolling invariably gets around to all the necessary bits and pieces that make life easier and trolling more productive. To refer to these items as accessories is really not giving them their just desserts. Most of the following items are really necessities for modern trolling. What you end up with on your own boat will be a matter of how you want to fish, availability, and how far your budget will stretch.

ROD HOLDERS

Storage in most small boats can often be a real dilemma. Some boats are well enough designed and set up to accommodate both horizontal and vertical rod storage with horizontal storage under the gunnels (or in the floor) and vertical storage next to seats or consoles. Whatever storage you have, make sure that you can fix your rods and reels down to avoid the hassle of having them bouncing around the boat, which usually results in broken rods. If you think about how and where rods usually get broken (the tips), it pays to put some thought into how you'll keep rods stored in your boat. In my own boat I have both horizontal and vertical storage for rods not in use. My horizontal storage has individual tubes to protect delicate rod tips; ensuring that you have this kind of set up is ultimately less expensive than breaking an expensive rod.

A good starting point for checking your boat set up for trolling is to look closely at your rod holders. Unless you plan to do all your trolling by hand holding a single rod, you'll need good rod holders. The tube style rod holders (which are usually found near the transom or in other awkward positions) that many boat manufacturers supply as standard equipment on boats, might be suitable for rod storage but as rod holders for freshwater trolling they would make a good anchor!

Look for rod holders that will accommodate the style of rod butt that you prefer and are easy to get

Freshwater Trolling

the rod out of when a fish strikes. I prefer rod holders that can rotate a full 360º and be tilted up and down for the maximum versatility. Having a rod for trolling sticking up in the air as it does in a tube style rod holder is a recipe for hassles like tangles and lures not reaching their maximum depth. There are a number of good rod holders available from Scotty, Attwood (Roberts) Fish On, Danica, Tite-Lok and a host of others. The cost of rod holders won't break the bank, but can really make a difference to your success.

Scotty Rod Holder

Attwood Rod Holder

TACKLE BOXES OR BAGS

Storage for items like tackle boxes or tackle bags needs careful consideration. If you happen to have the attitude that many lure anglers have, namely, that you can never have too many lures, you need a means of accommodating all these valuables.

The new generation of soft tackle bags and some hard tackle boxes from companies such as Plano can give you a lot of flexibility. These bags and boxes have removable plastic containers that can hold items including lures and terminal tackle. If you happen to spend time chasing a few different species this means you can have lures for each species in a separate container (or containers if you are a real lure junkie like me) and simply swap over the containers you need on the day.

If you look closely at your boat's storage space it may also be possible to have a built-in space to put the containers and not have to use the bag in the boat. Anything that you can do to eliminate clutter or objects to trip over will make your fishing experience that much better!

Built in storage for tackle boxes can help eliminate problems created by gear rolling around the floor of your boat!

Plano Tackle Box with removable containers.

Electric Motors

The popularity and availability of electric trolling motors has added a new dimension to trolling. Models available can be bow mount, transom mount or fixed to the main motor.

The development of electric motors for marine applications has been a real bonus for freshwater anglers. If you happen to own an older two-stroke engine or run a large outboard, attaining speeds slow enough to troll some of our big lures for cod can be a real challenge. Not only do electric motors provide the flexibility for cast and retrieve anglers to manoeuvre and position themselves for optimum casting, they can also be a valuable aid for trolling.

When you choose an electric for your boat, check the manufacturer's specifications to help determine how much power you need. Generally the heavier your boat (don't forget to include the weight of all your gear) the more power (thrust) you'll need. If you are likely to be fishing in windy conditions or strong current this usually demands more power and will also drain your battery of power more quickly.

Weigh up carefully how and where you will mount your electric motor. If you want the flexibility of using the motor for trolling and manoeuvring for casting, a bow mount motor is probably your best bet. These units are also much better at handling windy conditions.

Whether you are setting up a new boat or outfitting an existing craft, consider carefully the impact that using an electric motor for trolling will have on your boat set up. The quietness and minimal pollution of modern electric motors has to be weighed against the added weight and loss of storage space to accommodate batteries. With some batteries weighing 30–40 kg, where you decide to have your battery or batteries located can affect how your boat sits in the water and, on smaller boats, even affect how the boat gets up on the plane.

Drogues (sea anchors)

Known by a few different names, sea anchors or drogues are a simple means of helping slow down a boat when trolling or drifting. These units are generally made from material like nylon or other ripstop fabrics and come in shapes like a huge funnel or a bag shape. When deployed from the boat the sea anchor opens up like a big bag or umbrella and creates drag to slow the boat. This device can be very useful if you are attempting to troll with a strong following wind or if you have a large outboard that will not troll slow enough for the species you are targeting.

Sea anchors are generally fixed to the gunnels or bow of your boat and often one is adequate to slow the boat down. Be aware that running one anchor may cause the boat to veer in one direction; it may be more useful to run two smaller drogues to achieve the desired result and to ensure that you can still steer the boat in the desired path. Whatever style of drogue you choose to run bear in mind that you need to match the anchor to the speed you want to achieve.

Drogues or sea anchors can give the troller the ability to troll very slowly, even with large outboard engines.

Trolling Baffles

Trolling baffles or trolling plates are valuable tools for many trolling applications. If you own an older two-stroke motor or have a large motor that you want to use for trolling, chances are you may find that you can't get your speed slow enough. Trolling baffles generally have a spring-loaded mechanism to trip the plate and allow it to lower over the propeller; a pull chord attached to the plate usually

Trolling baffles such as the Happy Troller (from Magnum Fishing Products) can give the angler the ability to troll dead slow even with big outboards.

controls this mechanism. Most new four stroke marine engines are capable of trolling at very slow speeds without any undue damage to the engine, but you may still need to use a baffle if you want to spend time chasing Murray cod or other native fish species.

Trolling Boards

An easy method of fixing downriggers and rod holders to your boat and maintaining some flexibility is to use a trolling board. These boards can be made simply with timber or aluminium and fixed to a rail, pedestal or directly to the gunwales or side decks. This type of mounting board is often a very versatile means for mounting multiple downriggers or accommodating a set up in a small boat. If you use your boat for a range of fishing applications, rod holders and downriggers can often get in the way. A trolling board can give you the flexibility to move all this gear out of the way with a minimum of fuss.

Temperature, Depth and Light Intensity

No matter what type of trolling you do, whether shallow water or deep downrigging, knowing the preferred water temperature for the fish species you are targeting can be critical to your success. Most modern sonar units come with an option of speed and temperature that give you both surface speed and water temperature; helpful, but not the full picture. The Color-C-Lector units that were available some

Trolling Boards such as this angler made version can provide a great deal of versatility when setting up your boat.

Boats for Freshwater Trolling

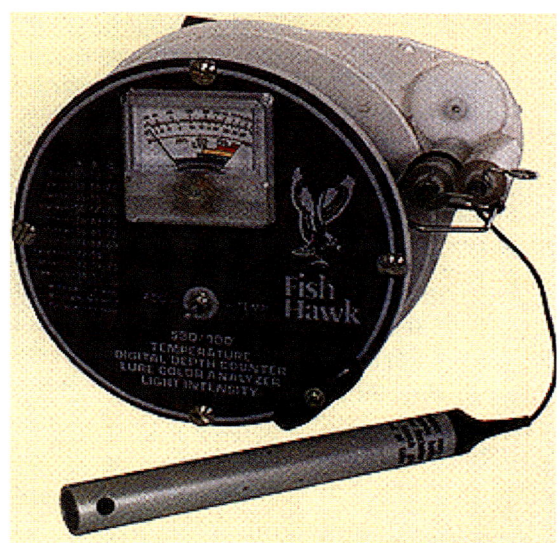

The Fishhawk model 530 pictured is a compact unit that incorporates a reel with attached cable and probe that is lowered in the water to determine accurate digital readings of depth, water temperature and light intensity.

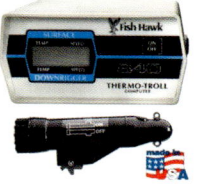

A Speed and Temperature unit, though not always readily available in Australia, is an extremely valuable item for serious downrigging enthusiasts.

home in on the preferred temperature of any fish species.

Another option for speed indication is the Luhr Jensen Trolling Speed Indicator. The indicator consists of a dial marked with speeds and a pointer attached to a cable and weight. To use this unit it is fixed onto a rail or gunnel and the weight is lowered into the water. Water pressure against the weight moves the indicator on the dial and gives you a reading of speed. The thing I like about this unit is its ability to provide you with an accurate and consistent depiction of your trolling speed. Because the weight is suspended in the water, and not sitting on the surface, the reading on the indicator is not affected by wave action or wind like a paddle wheel indicator. When using this indicator I often make a mark with a felt tip marker on the dial when I have established a speed that is effective for a particular lure.

years ago consisted of a large dial with an attached probe on a cable that gave an indication of light intensity (colour) temperature and ph levels are no longer being produced. Fishhawk (USA) produces a similar unit that displays digital depth, temperature and light intensity for lure colour selection. These sophisticated little units take the guesswork out of establishing water temperatures at the surface or depth.

Speed and Temperature Units

Downrigger speed and temperature units although relatively unknown in Australia, are a very useful device for trolling. These units will tell you both speed and temperature at the depth you have your downrigger set to, as well as on the surface. There are three major overseas manufacturers of speed and temperature units that I have personally used, the Cannon Speed-N-Temp, Moor Electronics Sub Troll 900 and the Fishhawk Thermo-Troll 840 and 920. These units share some features in that they all use a separate sending unit that attaches to the downrigger cable or bomb to send a signal back to the monitor where you get a digital readout. The big advantage of these units is that they give you valuable information on what is happening down where your trolled lure or bait is. Any fluctuations in current or temperature allow the angler to

The Luhr Jensen Trolling Speed Indicator is a great tool for judging speed while trolling

Lure Retrievers

No matter how competent an angler you happen to be, or how careful you may be with your lures when trolling, inevitably you will get your lure or presentation snagged on some underwater obstacle. This is doubly true if you fish heavy structure for our native fish. If you are fishing open water for trout or salmon and happen to have four lines out, then it's up to you to decide if it's more expedient to lose valuable fishing time pulling all your lines out or to sacrifice a lure. With many lures appropriate for trolling costing $15.00 each, you often have to make the choice of attempting to retrieve the lure or continue fishing and sacrifice the lure that's snagged—not always an easy decision if you're in the middle of a hot patch of fish. Thankfully there is now a range of good lure retrievers available, both imported and locally made. Although these items are not part of the boat they can make a big difference to your peace of mind and your costs.

An aerated Livewell tank might be a necessity if you plan to fish competitions or wish to keep fish alive for a period of time.

Lure Retrievers come in a variety of designs, usually incorporating lead and/or stainless steel and often chain to help get your lure back.

Livewell or Bait Tanks

The popularity of numerous fishing competitions and tournaments in Australia has had an impact on the available options on new boats. One such option is the aerated livewell tank. If you plan to fish competitions or tournaments on a regular basis this should be an essential item of equipment. Look carefully at the size of the unit; if you spend a lot of time chasing big cod a tank that is 600 mm long is not likely to be adequate. You can also make a good portable livewell from a large plastic container and an aerator. This system also has the advantage of being able to be removed when not in use and is inexpensive to make.

Travel Covers

Travel covers often seem like an expensive luxury for small-boat anglers, but I would suggest they are a very important addition to any boat. With many freshwater anglers travelling long distances on unmade or questionable roads it's sometimes easy for tackle and your other personnel belongings to bounce around in the back of the boat and actually fly out. I've known a surprisingly large number of anglers who have lost valuable gear this way. Travel covers also enable you to arrive at your destination with your gear clean and dry, even if you have been travelling dusty unmade roads.

Fitting out your boat for whatever style of fishing you do should be a rewarding experience, not some onerous chore. Avoid the rush to make permanent changes; drilling holes to fix accessories and all the bits and pieces should be the last thing you do. Take your time when making decisions about what goes where. If possible, fish from the boat before you start making a lot of decisions. Using the boat before you make changes will allow you to gain a better appreciation of how it suits your requirements and ultimately you will end up with a better result.

CHAPTER 2

Locating Fish

As anglers we all face many of the same dilemmas, regardless of the style of fishing we choose. Bait fishing, trolling, casting, spinning and fly fishing all rely on two main concerns: namely finding fish and getting them to bite. No matter what type of fishing you pursue, locating fish has got to be one of the most important facets of sport fishing. If you fish from a boat a depth sounder or sonar (short for Sound Navigation Ranging) is a vital piece of equipment. In addition to this equipment, the challenge of learning all one can about a fish species and catching these fish is an important factor in why many of us take up the sport. Rather than talk about choice of lure fly or bait I'd like to concentrate on locating fish including using sonar.

Each fish species reacts differently to their environment. As a result of both habitat and to some extent genetics, fish react to stimuli through their senses: sight, smell, taste, touch and hearing all play a part in how any fish species will react to your presentation. Locating fish also depends on a specific species' need for food, comfort and the biological need for reproduction.

To satisfy their needs and instinct for survival different species of fish have different requirements or preferences for food. Factors such as temperature, weather conditions, and time of year or season can all influence how fish will react in their chosen environment. Spawning requirements vary considerably from one species to another, as each species is very different. Developing an understanding of different species is vital to becoming a successful angler. To catch fish consistently, targeting a specific species requires you treat each species differently and to fish for each species differently. In both salt and fresh water,

Locating and catching fish such as this healthy rainbow trout requires a lot more than just good luck. Modern depth sounders are vital to improving your success.

structure plays a huge part in any species preferred habitat.

Fish located in dark stained water are generally more conscious of vibration due to the lack of visibility and usually rely on a mixture of sight, touch and hearing to locate their food, while fish located in very clear waters are likely to be more sight oriented. Pockets or areas of weed are always good spots to target fish. In most of our impoundments these areas can provide fish with yabbies, mudeyes, invertebrates and a host of other insect life. Areas such as river mouths or any area where water comes into or leaves a body of water, has an impact on fish location.

Interaction or the relationship of how different species behave toward one another also has a big influence on fish location. Available food sources, fish numbers and competition between species play an important role in fish location. Some species of fish are compatible, while others are not. Available food sources determine growth rates of individual fish and the general health of a fish population. A range of outside factors also directly influences the survival of any fish population. This can include factors like localised weather conditions, seasonal fluctuations in temperature and water levels as well as any intervention by human beings. Intervention by man in the form of boats, fishing pressure, pollution, swimming, artificial or other changes in habitat all affect fish location and ultimately, our fish catches.

Although some of the above mentioned conditions are beyond our control as anglers, many of these mysteries of underwater conditions can be unravelled with the help of modern sonar. In the past few years the sophistication of electronics and computer technology has leapt ahead. The application of this technology to sonar and GPS has resulted in a real bonus for anglers. The latest generation of sonar can show you an entirely new view of the underwater world! Over the years I've used a lot of different sounders, most of which performed admirably, but my new Lowrance LCX15 is one truly remarkable piece of equipment. The power and clarity of this unit has to be seen to be believed.

Water Clarity

Locating fish and their food sources without the aid of sonar can be a tall order. Using the LCX15 for

Native fish, such as this golden perch, are prime targets for anglers fishing our impoundments.

this task will allow you to target schools of baitfish and weed beds that are likely to hold food for a hungry predator. When searching likely locations for fish it pays to keep in the back of your mind that you are looking for habitat that fish are comfortable in, but you also need to consider the environment for their prey or forage. Water temperature and clarity plays an important role in any fish behaviour or activity. The type of water you fish also has an effect on the operation of your sonar. Sound waves from your sonar travel easily in clear fresh water. Wind, currents, suspended particles and micro-organisms also have an effect on the signal in freshwater, but this is generally minimal when compared to saltwater. In salt water, sound is absorbed and reflected by suspended material in the water and requires the use of lower frequencies to remain effective. The LCX15 has the ability to run dual frequencies—very handy if you fish in both fresh and salt water.

WATER TEMPERATURE

Water temperature, both surface and at depth, play an important role in fish behaviour. Finding the preferred temperature range for any given species is always a good starting point to locate fish. Keep in mind that each body of water or fishing location will be different when considering water temperature.

Chart: Preferred temperatures for Australian freshwater species.		
	Suitable range °C	Most active °C
Brown trout	3–25	13–20
Rainbow trout	3–25	10–18
Chinook salmon	3–25	10–18
Murray cod	6–26	18–22
Golden perch	4–28	14–26
Silver perch	4–28	14–26
Macquarie perch	4–28	10–17

Water temperature has a direct influence on spawning fish but also plays a big role in the daily patterns of any fish species. As little as 2 degrees in temperature can often spell the difference between fish in the net and failure. With the numbers of temperature probes available today there is no reason why every angler can't monitor temperatures at almost any depth.

The temperature in any lake or impoundment is rarely the same from the surface to the bottom. Reams of information have been published by many scientists about stratification, the layering of warm and cool water in lakes and impoundments. For anglers, the important part of this process is where the two layers meet: the thermocline. The depth and thickness of the thermocline can vary with the time of day, season, intensity of sunlight and wind or wave action. The thermocline is never an exact band, but rather one that is constantly changing in depth, intensity and size. In deep lakes the water may even stratify in to two distinct thermoclines. The band of water, which forms the thermocline, is important because many species of fish like to suspend in, just above or just below the thermocline.

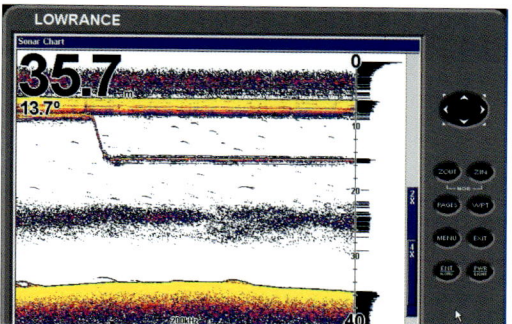

Figure 2.1: Sonar image of fish suspended near a thermocline.

Have a look at Figure 2.1 which depicts an impoundment, in this case Moogerah Dam in South East Queensland. This clearly shows two distinct thermoclines. Many species of baitfish will suspend just above the thermocline with the larger predators in or slightly above it. Monitoring the location of the thermocline is easy with the aid of sonar. The larger the temperature difference between the layers of water the more dense the thermocline shows on the screen.

STRUCTURE

Structure in any lake or body of water can encompass a range of factors. When referring to structure it is necessary to consider the range of structures in a species' environment. Generally most anglers think of structure in terms of submerged trees, rocks, shipwrecks, etc. Structure also includes elements like bottom configuration, bottom contents, water movement, vegetation, drop offs, submerged points, and water depth.

When prospecting a new area or body of water, watch your sounder carefully. A great looking

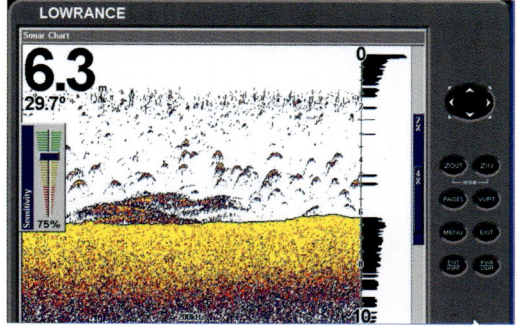

Figure 2.2: Locating structure is vital to locating most species in our freshwater impoundments.

Targeting our native species such as this Murray cod in many of our impoundments has become one of the fastest growing pastimes in the freshwater fishing scene.

bit of fish-holding structure near the bottom of a lake that stratifies in summer is likely to be a poor target area if it is below the thermocline and has poor oxygen levels in summer. The depth of any structure will determine how fish use it. Extensive weed beds in very shallow clear water are unlikely to be frequented by predators in intense summer daylight. Any structure used by fish needs to be at a depth that is appropriate for a given species' requirements. Bottom content including rock base, weed beds, boulders, sand or mud can all make a difference to your success. Don't waste your time pursuing an area that doesn't have the right bottom structure to attract the species you are targeting. The sonar image in Figure 2.2 depicts a classic example of fish seeking structure. Don't get caught in the trap that most anglers fall for, structure or the shape of an area alone don't attract fish. The best fish-attracting areas generally show a combination of structure, weed, depth and available food. Fish often make use of different areas according to the season or time of year. To be a really successful angler means not only using the right techniques and gear, but also developing an understanding of fish species and their environment.

CHAPTER 3
Lures for Trolling
Salmon and Trout — Golden Perch and Murray Cod

Colour Selection for Lures

Many of our impoundments contain an abundant supply of galaxids and these minnows are a large component in the diet of the trout and salmon stocked in these lakes. But even though many anglers attempt to 'match the hatch' by using colours or finishes on lures to match the existing forage or baitfish this often is not successful. Matching the hatch in this instance probably has more to do with size and shape of a presentation than colour. Lure colour selection should be based on what our target species is likely to see at a particular depth. I would suggest that colour choices should be based on a combination of water clarity, light conditions and water temperature.

Choosing lure colours for effective trolling can be a fairly confusing task if you rely on the information on lure packages, fashion, or anecdotal information from our fishing mates. Let's face it, we've all heard the saying that lure colours and finishes are designed to catch fishermen not fish.

Unlike casting and retrieving lures, trolled lures are almost always worked out of sight. To ensure success you must know what your lures are capable of, including diving and running depths.

Whether it's a marketing technique to sell lures or a genuine attempt to create finishes and colours that actually do work to catch fish, colour is an important part of lure selection. There has been a wealth of scientific information published to support the notion that fish do distinguish between colours. Not being a scientist much of this information has probably been wasted on me, with perhaps two major exceptions. Paul Johnson's book 'The Scientific Angler' and Dr Colin Kageyama's book 'What Fish See' are both loaded with information and the latter especially written for anglers in a clear easy to understand style.

Much of the information that we get from advertising, the media and writers is not always as clear and accurate as it should be when it comes to the description of colours and finishes. Terms and descriptions like 'bright', 'glow in the dark' 'fluorescent' etc, all get tossed around with little or no understanding of what they really mean. I would suggest that when we choose colours for lures (or flies for that matter) most anglers are thinking in terms of what colour a lure is in air, not what colour it will be in the water at a given depth. What we really need to consider and understand is what happens to our lure colours under the water. The list below provides a few working definitions for some common descriptions when talking about colour.

According to most optometrists objects in air do not catch your attention based on colour, most catch your attention because of brightness or movement. In the air light comes from certain directions either from artificial sources or sunlight and its direction is easy to determine because it casts a shadow. Water is however, very different, once you are more than 1–2 m under water the water appears to glow and light seems to come from all directions. As light travels through water it is reflected and scattered. The more dust and material suspended in the water, the more the light is scattered. When a fish (or human diver) looks at distance through water the scattered light takes on a uniform glow. You can check out this phenomenon yourself in your swimming pool, or a clear river or lake.

For trolling anglers (and most other forms of fishing), long distance colour shifts can play a big part in your success or failure. Trolling usually means that your lure is passing by a fish fairly quickly with little or no chance for making repetitive presentations. In this instance fish are usually striking out of hunger and/or reflective action. A fish's ability to hear and see a prospective meal is an important part of foraging and survival. Let's assume that in clear green water a fish is able to see your lure from 7 m away, it makes the choice to go for your lure and charges the lure. At 1 m away from the lure this fish puts on the brakes and shies away from your presentation. Why? The lure you were trolling was bright red, but at a distance of 7 m the lure appeared black. The colour shift from black to red was probably unlike anything the fish had ever experienced in nature. If the lure had maintained its colour over the entire distance it's a good bet that this fish would have taken the lure.

Colour shifts occur as water filters out different wavelengths of light depending on the colour of water and the type of suspended material in the water. Clear blue, green and brown water all filter out light in different ways. As light passes deeper in each of these water types both short and long wavelength light waves are filtered out. Light penetrating in deep water becomes monochromatic or one coloured. In other words the light penetrating deep blue water becomes blue, deep green water becomes green light etc. When light penetrates deep water and becomes one coloured, the only lure colours that will remain bright, are those that match the colour of the water, are white, or are fluorescent

---

LIST OF DEFINITIONS

Developing an understanding of how the colour of artificial lures changes under water can help us all become more successful anglers. To improve our knowledge of colour a few basic definitions are helpful, such as:

- **Light:** a type of radiation that can be detected by eyes. Light travels in 'waves' of different length. Short wavelength light includes ultraviolet, purple and blue. Medium wavelength light includes green, chartreuse, and yellow. Long wavelength light includes orange and red.
- **Fluorescence:** Is the ability of an object to reflect light of a longer wavelength than it received.
- **Brightness:** Is the ability of an object to reflect a large amount of light.
- **Phosphorescence:** material that continues to shine in the dark after exposure to light; 'glow in the dark'.
- **White Light:** is a light that is a combination of all visible colours. This type of light includes radiation from a variety of different wavelengths or colours. In order for light to appear white, it must include blue, green, yellow, orange and red components.
- **Colour shift:** the phenomenon of objects apparently changing colour under water due to the light filtering nature of water.

The range of colours, sizes and running depths for lures is almost limitless. Most freshwater trolling specialists tend to have a broad range of lures to suit prevailing conditions.

colours of a longer wavelength. All other colours will turn a non-descript dark colour.

For anglers of all methods, the range of colours and finishes we have to choose from is amazing. Colour and subsequent visibility are crucial to all successful lure fishing. These factors have a big influence in attracting fish to your presentation, but they alone are not the only reason a fish will attack a lure. Factors like a lure's action and speed, the water's clarity and temperature also have an impact on a lure's success. We may never know conclusively which one of these factors is the most important in a fish's selection process for its next meal. In the meantime, I for one will continue to treat all of these factors as vital to my lure fishing success.

Lure Design

The main factors in lure design affecting lure depth include the lure's profile, buoyancy, action and lip or bib design. These elements are what determine any style of lure's depth profile or range. The single most important design factor affecting depth of any diving lure is the surface area and shape of the lip or bib, along with weight, which largely determines how deep a lure will run at a specific line out length. With a bit of experience we can get a general idea of how deep a lure will run by its overall appearance. This is a very hit-and-miss process: it is very easy to over or underestimate depth by this approach. There is an easier way to get a more accurate result. Freshwater Fishing Magazine's publication, 'Frank Prokop's Lure Encyclopaedia' and 'Precision Trolling' by Dr Steven Holt, Mark Romanack and Tom Irwin, are two excellent sources of information. Between them, these publications cover a broad range of lures both local and imported. Another alternative to get accurate depths is to troll your lure at your desired line out length and get someone with a very good depth-sounder, such as a Lowrance LCX 15 or LCX 19 to follow and record what depth the lure is running. These sonar units are more than powerful enough to pick up a trolled lure and give you an accurate depth. It may seem like a lot of messing about, but you can easily check the running depth of a lot of lures in an hour. Well worth the results in the long term!

TUNING

A critical area of trolling that often gets overlooked in the rush to get into a fish is the tuning of lures. A correctly tuned lure should dig straight down and move forward in a straight line with the boat. Badly tuned lures on a long trolled flatline will track across other trolled lines creating havoc by tangling the other lines. If the lure happens to flip over it

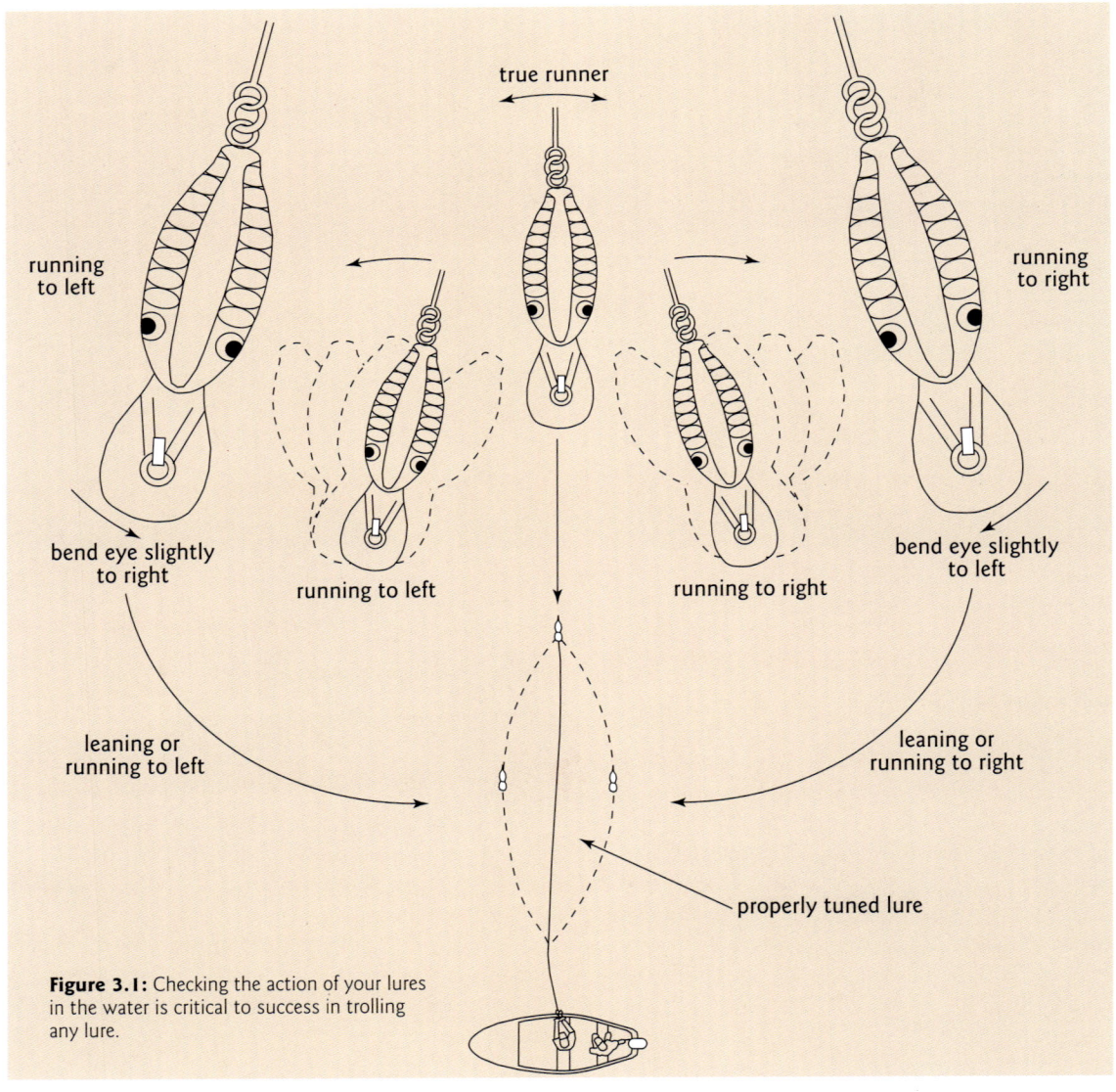

Figure 3.1: Checking the action of your lures in the water is critical to success in trolling any lure.

can also cause severe line twist. Always check your lure's tuning at the start of every troll. Make sure you check the lure at the speed you intend to troll. If it tracks to one side, bend the nose loop in the opposite direction. Most good quality lures will come properly tuned straight out of the box, but it always pays to check!

Lures for Flatline Trolling

Lures for flatlining come in all shapes and sizes with a range of actions and diving capabilities. Lures with different running depths can be used to target fish in different parts of the water column. You can run a lure right on top of a shallow rocky bottom with a shallow running floating lure or fish an 8 m drop off with a deep diver. The key to successful trolling of any lure lies in understanding each model's properties, including optimum speed, depth capabilities and action. The following text and graphics will hopefully give you a few insights into making your lures more productive.

Minnow Style Lures (Floating)

The original floating Rapala would have to be the all time classic in this style of lure. Slim minnow profile lures abound, with some fantastic finishes available. Imported lures that are proven fish takers include Storm, Rebel, Bomber, Team Daiwa, Yo-zuri, and Lucky Craft. Australian producers Predatek (MinMin), Legend, StumpJumper, Halco, Oar-Gee and Australian Crafted all produce excellent fish catching products.

Jointed Lures

These are normally floating lures, though there are sinking models available. Rapala Jointed, Rebel

Jointed minnow style lures, with their seductive swaying action, are another proven producer for trolling applications.

Broken-Back Minnows, Nilsmaster Jointed and the Tasmanian produced Wals Lure are effective versions for trolling. The jointed lure is essentially the basic floating version, separated into front and rear halves, which are connected by freely moving, interlocked metal rings. The result is a lure that tracks and trolls in the same manner as a straight lure, but with a wider, wilder wiggle as the two halves swing back and forth. This style of lure is a great alternative when standard lures are not producing; I've had success with all of these lures under varying conditions, and they remain one of my all time favourites.

Suspending Minnows

This style of lure is one of the more 'high-tech' offerings by a range of manufacturers, including Rapala, Storm, Halco and a host of others. The Rapala Husky Jerk breaks the company's long-standing tradition of balsa-wood lures with a moulded plastic body. This model neither floats nor sinks, its neutral buoyancy causing it to suspend

Floating hard-bodied minnow style lures are one of the all-time best performers for trolling applications.

Freshwater Trolling

Suspending lures have become a popular option for most cast and retrieve lure anglers. They can also be a valuable tool for the trolling angler.

Sinking or countdown minnows are very productive lures for trolling. This style of lure needs to be let out under tension to avoid hanging up on the bottom as you release line from your reel.

when stopped at whatever depth its lip has pulled it to during a retrieve. Trolled, it has characteristics similar to floating lures, with some key differences. The additional weight required to create a suspending lure causes it to run slightly deeper than a floater of the same size, and it has a slightly wider wobble as well.

It also has a bit more flash and shine in its finish than painted, balsa lures, and contains metal balls in an internal chamber that give it a loud, fish-attracting rattle. This type of lure tends to stay tuned longer and can take more abuse than a wooden model, and the 'bells-and-whistles' really do serve to give an edge when fish are not striking a standard minnow lure. In the copper/black, rainbow trout, and baby bass finishes, this lure has accounted for some great fish.

Sinking/Countdown Minnows

Yes, you certainly can 'count-down' this sinking lure, casting it out and letting it drop, at one foot per second, then retrieving it at a desired depth. It is also a deadly trolling lure, with the larger sizes accounting for many trophy trout over the years. This weighted lure needs higher trolling speeds in a given size than the same size floater to achieve optimum action, and unlike with floating lures, the trolling line must be let out under tension, so the Countdown does not sink to the bottom and hang up. Its ability to run deeper, and to drop rather than rise during turns, gives it a unique action, producing fish that would not move upward to strike a floating lure. This style of lure in the 9 cm or even larger models is often very productive when trolled along steep shorelines or drop offs.

Deep Divers

Most deep diving lures are floating lures, but whatever their weight, what differentiates this type of lure is its oversize plastic bib, which digs it deeper per weight and body size than a lure with a smaller lip. A major advantage of large-lipped, floating divers is their split personality. They float at rest,

Deep diving lures (such as the models pictured) provide an easy means of targeting fish in the 2–6 m depth range. Some of the locally produced lures are capable of being trolled to depths of 10 metres.

yet dive deep on the retrieve, deeper still on a long-line troll. Letting out a trolling line is therefore as easy as setting your reel on free-spool, and if the lure runs too close to the snags and starts to hit bottom while trolling, it can be allowed to rise shallower by slowing or stopping your boat, avoiding hang-ups.

There are a vast array of deep diving lures that are suitable for trout trolling, with a range of profiles and actions. Australian producers like MinMin, (Predatek) Bennett, McGrath, Knols, StumpJumper, Legend Helmax, Oar-Gee and Australian Crafted to name but a few. The imported lures that I favour include brands such as Nilsmaster, Rapala, Storm, Team Daiwa, and Lucky Craft. One of my favourite deep divers in a minnow profile is the Storm Deep Jr. Thunderstick, which will troll at nearly 6 m on 0.22 mm line.

Winged Lures

Without doubt the most popular and easily identifiable lure style for trout fishing in Australia would have to be the Cobra or Tassie style of lure. Manufactured by Wigstons, Lofty's, Tillans etc., they are available in a range of sizes, diving depths and colours, with the applications for flatline trolling this lure being almost endless. Changing speed and lure size can give you the ability to cover depths from 0.5–3.5 m on a trolled flatline.

The standard cobra style lure (12.5–13 g) has been the mainstay of trout fishing in Australia for a lot of years, because they catch fish! This size lure will troll at approximately 1.2–1.5 m on a flatline, depending on speed and line diameter. Trolling speed for this lure is within the range of 0.5–3.5 km/h. You can easily check the action of this lure by watching your rod tip. A regular rhythmic pulsing action will insure that the lure is working properly.

If you asked most Australian trout anglers if they ever used spoons for their fishing most would likely reply that they seldom ever use this type of lure. In reality the Tasmanian 'Cobra' style of lure is really a type of spoon, albeit a heavy, uniquely-shaped lure, it is still basically a spoon. Every size, shape and description of spoon has been manufactured over the years, but nothing else comes close to these little plastic and lead marvels. The Cobra style of lure has an amazing scope to accommodate a broad range of applications for almost any fishing conditions.

The Cobra style of lure has been around for a

The array of 'Cobra' style lures pictured is a small part of the overall number of colours available for this style of lure.

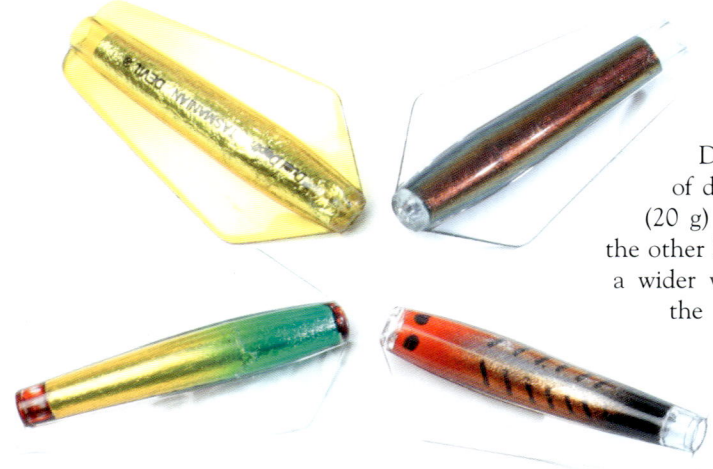

Two new options for the Cobra style lure include the Dual Depth Tasmanian Devil and Lofty's Wide Wing Cobra.

The introduction of the new Wide Wing Cobra and the Dual Depth Tassie Devil should prove to be a real bonus for trout anglers in the coming seasons. Wigston's Dual Depth Tassie Devil uses a combination of dual towing points and heavier weight (20 g) to achieve a deeper diving lure. On the other hand Lofty's Wide Wing Cobra adopts a wider wing and retains the 13 g weight of the standard Cobra. My experience with these lures has shown that they are both fairly sensitive to speed. The Dual Depth needs a slightly faster (2.5–3.5 km/h) because of its heavier weight while the Wide Wing seems to operate best at slower (1.5–2.5 km/h) speeds. Both lures need to run at the right speed to avoid line twist. As with any lure of this style watching the rod tip can tell you what's going on with the lure. If you pick up weed on the lure it will change the action of the rod tip, alerting you to the fact that it's time to bring the lure in to clear it. Both of these lures work equally well for lead lines, flatlines or for downrigging. One of the major considerations with these lures is to be aware that the action they have is different to the standard lure. This difference in action can often spell fish in the boat.

The following notes and diagrams about this style of lure should help create a better understanding of how this lure works and what you can do to improve your success and catch rate. There are a number of factors that play a part in how well your lure is presented to a fish including your choice of line, rigging, hooks, boat speed and the lure itself.

very long time now, with a range of manufacturers producing this type of lure. Until recently there has been little change in this style of lure, with most makers offering a range of colours and finishes, but remarkably most of the differences are little more than cosmetic. With all the ardent followers of this style of lure around, I am probably going to get hate mail for saying this, but the biggest difference in most of these lures has been the range of colours available. The design of lures with slightly different weights or subtle changes in the body size or wings has been the norm, but basically the lure has remained much the same, until now.

Two manufacturers, Lofty's Lures and Wigston's Lures have both introduced a new deeper diving lure. Although both these lure-making firms have no doubt been researching this problem for some time it's very interesting to note that their approach and the finished products are very different. Because of this the two lures, Lofty's Wide Wing Cobra and Wigston's Dual Depth Tasmanian Devil, both have a unique action and a range of applications. No matter what your personal preference in brand may be I would suggest that both styles of lure deserve a place in your tackle box to accommodate a range of conditions.

Rigging Cobra Style Lures

One tried and true approach to rigging this style of lure usually entails passing the line through the body of the lure, through a small plastic bead, and tying it to your choice of hook. As the lure moves through the water the hook is pulled up

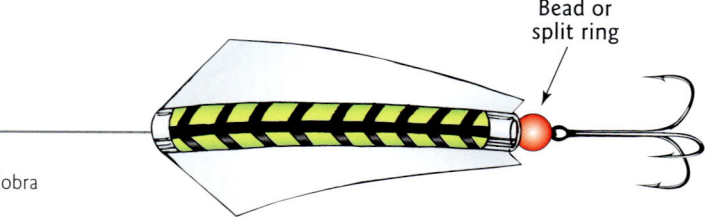

Figure 3.2: Rigging a Wide Wing Cobra

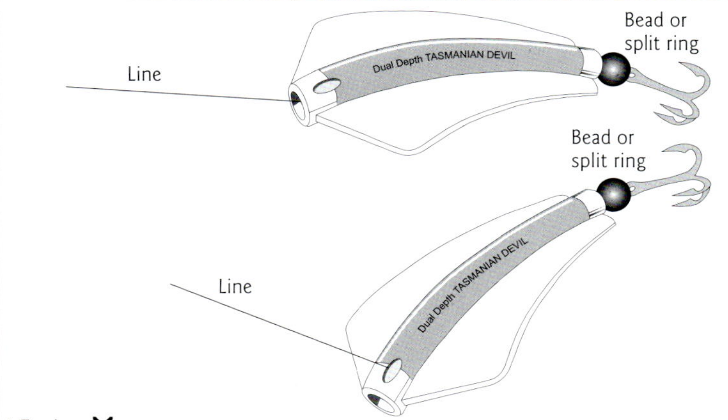

Lures for Trolling

Rigging the Dual Depth Tasmanian Devil in the standard mode entails running the line through the centre of the body, then through a bead and finally tying it onto the hook. This can be varied by passing the line through the body, then tying it directly to a split ring which has a hook attached to it. This allows the lure's action to be far more pronounced. Rigged in this method the Dual Depth Tasmanian Devil runs at 2.5 metres in depth.

Rigging in the deep running mode, the line is passed through the moulded hole in the nose of the lure, through the body, through the bead and then it is tied to the hook, or you can use the split ring method. Rigged in the deep running mode, the Dual Depth Tasmanian Devil reaches a depth of 3.6 metres. It must be noted that these depths can vary with line diameter and boat speed.

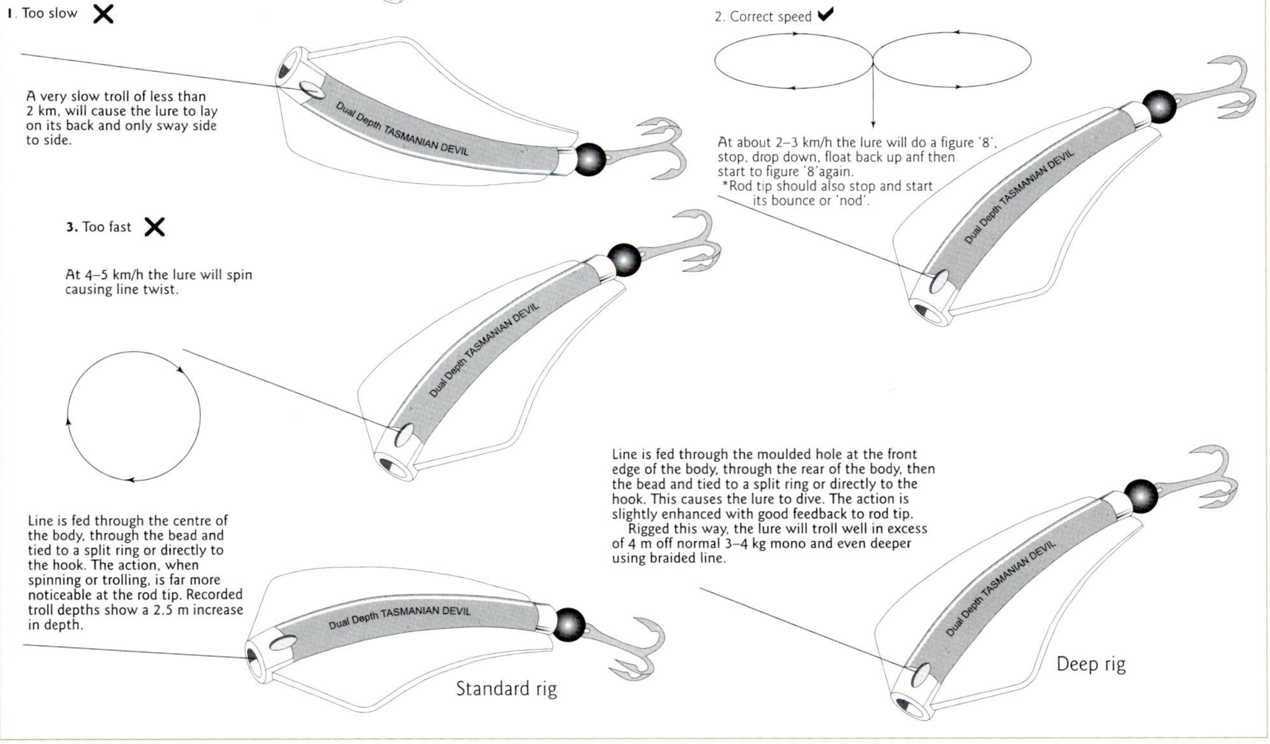

1. Too slow ✗

A very slow troll of less than 2 km, will cause the lure to lay on its back and only sway side to side.

2. Correct speed ✓

At about 2–3 km/h the lure will do a figure '8', stop, drop down, float back up anf then start to figure '8' again.
*Rod tip should also stop and start its bounce or 'nod'.

3. Too fast ✗

At 4–5 km/h the lure will spin causing line twist.

Line is fed through the centre of the body, through the bead and tied to a split ring or directly to the hook. The action, when spinning or trolling, is far more noticeable at the rod tip. Recorded troll depths show a 2.5 m increase in depth.

Standard rig

Line is fed through the moulded hole at the front edge of the body, through the rear of the body, then the bead and tied to a split ring or directly to the hook. This causes the lure to dive. The action is slightly enhanced with good feedback to rod tip.
Rigged this way, the lure will troll well in excess of 4 m off normal 3–4 kg mono and even deeper using braided line.

Deep rig

Figure 3.3: Rigging a Dual Depth Tasmanian Devil

tight against the bead, which in turn locks it in one position. To gain the most from your lure the hook should be able to move freely. Using a split ring instead of a bead, allows free movement of the hook.

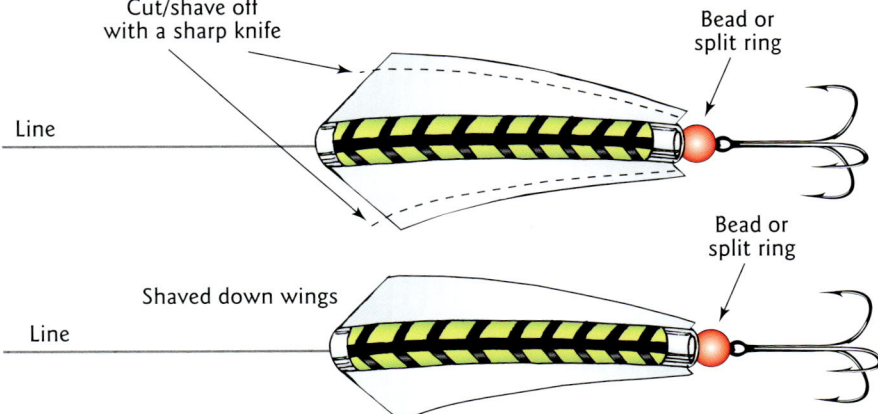

Figure 3.4: Rigging for shallow running

Hook Options

The standard hook that comes with most of Cobra style lures is usually a single treble hook. After experimenting with several different types of hooks for the last three or four years, I almost never use a treble behind a Cobra style lure. Invariably, if you have my sort of luck, when you hook up on a small fish they seem to have a real knack for getting caught with all three prongs of a treble in their mouth. When you try to get the treble out you can do a lot of damage to the fish. Catch-and-release fishing is something I firmly believe in and want to practise, so I now use single hooks. My favourite pattern is an open eye Siwash hook. An Eagle Claw 210NA, VMC 9171N1 or Gamakatsu Saltwater Fly Siwash hook are all brands I've used with great success. Gamakatsu have also recently released a new model of hook designed specifically for single hook lure applications. This hook has a short shank with a round open bend and large eye and should prove to be effective for this style of rigging.

One of my main concerns when changing over to single hooks was whether or not the hook-up rate would be as good as with a treble. My experience has been that it is even better and rarely do you seem to fail to hook a fish. The Siwash pattern of hook is used extensively in the US and Canada for salmon and steelhead. This type of hook features a very long point, short shank, and wide gape and usually

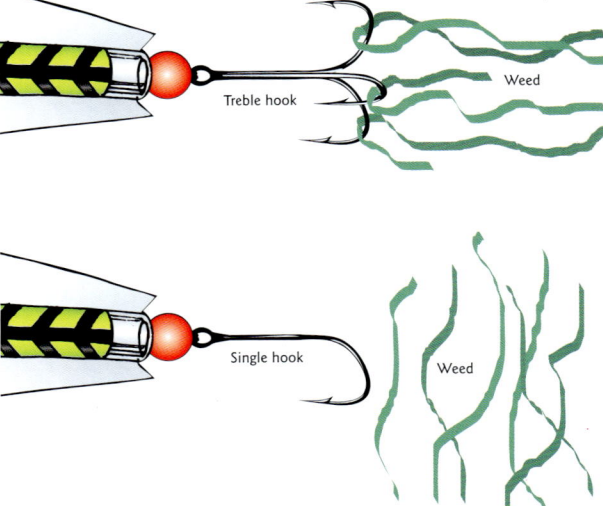

Figure 3.5: Single hooks versus treble hooks

has an open eye, which you can crimp over a split ring. Regardless of what brand or pattern of hook you choose, make sure you keep it razor sharp.

Most anglers seem to shy away from the 7 g or 25 g size lures for trolling but I've found they are both very productive. The size of the forage fish you are trying to imitate should dictate what size your lure should be. The 25 g lures have a very strong action that can be extremely productive if you are chasing big trout and salmon. The smaller 7 g size lure is great for shallow water or weed bed trolling and has the same great action. The 7 g model is also an excellent choice of lure when you want to run a slider off the downrigger.

The 7 g lure manufactured by Wigston's and Lofty's, is a much under-utilised option for trolling. This lure is often a great choice when you want to troll a shallow weed bed or just want a smaller presentation to offer fish that are feeding on small baitfish. Trolling on a flatline will get you down to approximately 0.5 metres.

The introduction some years ago of the Dual Depth Tasmanian Devil and Lofty's Wide Wing Cobra has given trollers more options for covering different depths while flatline trolling. Don't get caught in the trap that has happened with many anglers using this lure for the first time. The usual pulsing of the rod tip when trolling the standard lure is not what you get with this lure. What you need to look for is a stop-start action on the rod tip, caused by the lure doing a figure-of-eight movements,

The 7 g Cobra from Lofty's Lures and Wigston's are both excellent performers when a small presentation is the order of the day!

stopping and then starting again. The variation of action to this lure is one of the main reasons it can be so productive, unlike lures that have a constant even action. I've found the Dual Depth great for speeds up to 3.5 km/h, while the Wide Wing seems to handle the slower speeds better due, I suspect, to the different weight of the lures.

With the advent of the new deeper diving lures, the amazing range of colours and sizes means Australian anglers have a greater choice than ever before. Dollar for dollar these lures are some of the best value tackle around, so no matter which brand you choose you can support a unique Australian product!

THE SOFT OPTION

Soft plastic lures have finally come of age in Australia, though most anglers probably associate them more closely with fishing for natives, casting and retrieving, and jigging. Some of the new soft plastic products that are currently available are unbelievably realistic and well suited as an option for trolling. The minnow profile soft baits that I have had success with include Storm, McLaughlin's (Juro), Renowsky, Squidgies and Atomic. All of these companies have products with a range of great colours as well as prism and holographic finishes.

Another alternative for trolling is to use one of the new soft plastic swimbaits. Well known in the north American market as the hot producer for cast and retrieve applications when chasing bass, swimbaits are versatile lures that are effective for many species including barramundi, perch, cod and trout.

Although there doesn't appear to be a universal definition for 'swimbaits' within the tackle industry the term seems to encompass two main styles of lures. Minnow or shad shaped soft plastics are the norm, while the hybrid wood/plastic varieties such as the original AC Plug, a hard wooden body with soft plastic tail, also fall into this category. Many of the hybrid style lures also have a bib, which allows them to float at rest but dive when retrieved or trolled. Soft plastic swimbaits usually either float or sink at various rates depending on the plastic's formulation and the lead content. Storm lures have now also brought out a new suspending model, the Suspending WildEye Swim Shad.

Swimbaits generally have a wiggling tail action, but also gain action from the shape of the plastic body itself as well as the addition of a jig-head, if used. Many of the soft plastic models available also closely mimic the complicated action generated by bibed or diving lures known collectively as crankbaits. Swimbaits are available in a huge range of sizes from about 30 mm long to the monster catching models over 225 mm available from companies like Storm Lures and Juro. From boot or paddle tails to split tails and curl tails there are

A few of the many soft plastic lures available in Australia. Though mostly seen as cast and retrieve lures, they also work very well for trolling.

a host of swimbaits to cover almost any application or depth. Another big plus for swimbaits is their cost: you can easily buy two or three swimbaits for the cost of one hard body lure. When you add the ability to readily accept the new scents formulated for soft plastics it's easy to understand why this style of lure has so much potential.

Rigging Swimbaits

Some swimbaits come pre-rigged with a single large up-turned hook behind the head and some with both the single hook and a belly treble hook. If you use the single hook model it often pays to run a stinger hook to stop short takes. This can be rigged much like putting a stinger on a spinnerbait. Slip a treble over the point of the single hook and pin it in the back of the bait, then use a piece of tubing on the single hook to stop the treble from slipping. You can also rig a stinger with a short piece of braid or Dacron and tie it to the back of the single hook. The unrigged models of swimbaits can also be rigged to run weedless, by burying the hook in the soft plastic body, handy if you are flatline trolling in areas that contain a lot of snags and weed.

To troll swimbaits I have found that the medium (¼ ounce) weighted lures and suspending models work well for flatline trolling. Most of these models will troll to about 1.2 to 1.5 m on 0.22 mm monofilament and about 1.8 to 2.0 m on 10lb braid. Speed is fairly critical with the lighter models—too fast and they will roll. The heavier models can stand a fair bit of speed and can easily be trolled at a speed with Cobra style lures. Trolling swimbaits for trout has a number of things in common with trolling flies. This style of lure can greatly benefit from the addition of an attractor. For lead core line or downrigging with swimbaits the action from a small flasher or dodger can really improve your results. Attractors such as dodgers give this style of lure a lot more movement than the tail wiggle common to this style of lure.

To improve your results in low light and or very deep downrigging conditions, running an attractor off the bomb can be a great help. To further enhance the amount of light generated from an attractor there is a new product that I believe has a lot of applications for trolling. FireLights are a new product developed by Juro with a huge range of applications for both fresh and saltwater fishing. FireLights consist of a small coloured plastic cylinder (available in white, pink and chartreuse) approx 10 mm in diameter and 50 mm long. This small cylinder contains a water-activated, battery-powered strobe light. When you put the light in the water it automatically switches on and shuts off when you remove it from the water. Another advantage with this light is that it has a line guide, which enables it to be run on your fishing line. The applications for this little unit are endless, but in this scenario I like to rig it in front of the attractor run off my downrigger bomb. The amount of light that bounces off a set of bladed attractors or a large flasher is staggering.

Some of my favourite soft plastic lures for trolling include the Storm WildEye Swim Shad in both the rigged weighted model and suspending models and the small WildEye Finesse Minnow. Another range of swimbaits that I've had good success with is the Juro line of Firebaits. Excellent trolling lures include models from the Mini Wiggle and Mini Dog series, both of these lures make a great presentation

Swimbaits are available in a vast array of shapes and sizes.

The Juro Firelite is an easy way to get light to a lure or attractor that is trolled deep.

Rigging Dodger and Swimbait with Firelite: For added attraction run the Firelite in front of your attractor. To rig this set-up run a short trace of 300–400 mm of 15 kg braid, tie one end to the downrigger bomb, run it through the Firelite and tie the other end to your choice of attractor. Bladed Attractors, like a string of willow leaves or a large flasher, also work very well for this application.

in a 50 mm size bait. To troll a larger lure I would suggest you have a good look at the Sea Dog and Pelagic series from Juro. These lures have a slimmer minnow profile with some great holographic finishes and lifelike detail. To increase the action of this style of lure Juro have also developed their Sly Divers, a jig head with chemically sharpened hook and a diving bib to further increase the flexibility and applications of this style of lure.

Improving Lure Performance

Despite all that has been written in other publications, many anglers still persist in using lures straight out of the box. If chasing monster fish is your passion then you'll need every trick in the book to guarantee success. Tuning or tweaking to get the best from your lures is not really difficult, and often involves only minor alterations, such as changing hooks or painting stripes or eyes. Few new out of the box lures perform as well as lures that are tweaked by minor changes. With modern moulding and production techniques every lure is a little different. We've all seen this, inevitably one lure out of a dozen in your tackle box seems to out-fish the others. Each and every lure needs to be tuned for the best performance.

Minnow Style Lures

Minnow shaped lures, or stickbaits, as they're known in some parts of the world are some of the most effective lures on the planet. Whether you are trolling or casting these lures, they've probably accounted for more than their fair share of big fish. My all time favourite remains the floating Rapala. Out of the box these lures will catch heaps of fish, but a few minor alterations will help with the pursuit of big fish. The 7F (original floating), 9F and 11F Rapalas are all great fish catchers, but can do with a little help. Firstly, with the 11F, as with all stickbaits that have three hooks, remove the front treble—this will help with avoiding foul-hooking a small fish. If you've ever tried to land a fish that has one treble in its mouth then turns and tries to run and gets another stuck in its gills or body, you will know how it pulls like a gigantic spinner. Not a good scenario if you are planning to release the fish.

Virtually all stickbaits will benefit from small changes. With an 11F Rapala as an example, remove the two front hooks, leave the split ring on the forward hook eye to help keep the lure in balance. Replace the centre hook with a nickel plated size 6 treble. This minor alteration will change the action from a tight shimmy to a wider wobble and make the lure run truer; it also adds a little flash from the nickel hook. The larger centre hook also helps hook big fish more solidly. In this application, or with any other brand of lure, if removal of the front hook upsets the balance of the lure, you have a few options. If you don't want to remove the hook, just take a pair of pliers and bend the points in so the hook can't hook-up or use a larger split ring. Another option is to use lead wire wrapped around the split ring to balance the lure. Storm Suspend Dots or Strips, though primarily used to create suspending lures, also work great in this application. They are easy to apply and also come off easily if you have to make any changes.

SPOONS FOR TROLLING

In this age of high tech wizardry, using spoons probably sounds a bit mundane and downright old fashioned. Truth is, these thin bits of metal have some real advantages for trolling. Generally spoons are just about the least expensive of all the lures available to anglers today. The choice of finishes for spoons is absolutely staggering with plated finishes in gold, silver and copper, painted finishes including glow in the dark and every imaginable type of prism tape. Some spoons will catch fish straight out of the box, but the real secret to getting the most out of them is to tune them to cover a range of conditions and speeds. Tuning is often as simple as bending with a pair of pliers, changing hooks, adding tape or even simply changing snaps or snap swivels. I prefer to rig single Siwash hooks for this style of lure, and firmly believe they give a much better hook-up rate, as well as making it easier to release small fish.

Spoons for trolling should have an aggressive wobble not a lazy sway, and should be on the verge or rolling or spinning, with an occasional darting action. Lures that are very effective include offerings from Eppinger, Wonder (Flutterspoons), Stinger, Luhr Jensen and Sluk lures (these lures have a great finish). To tune a spoon use a pair

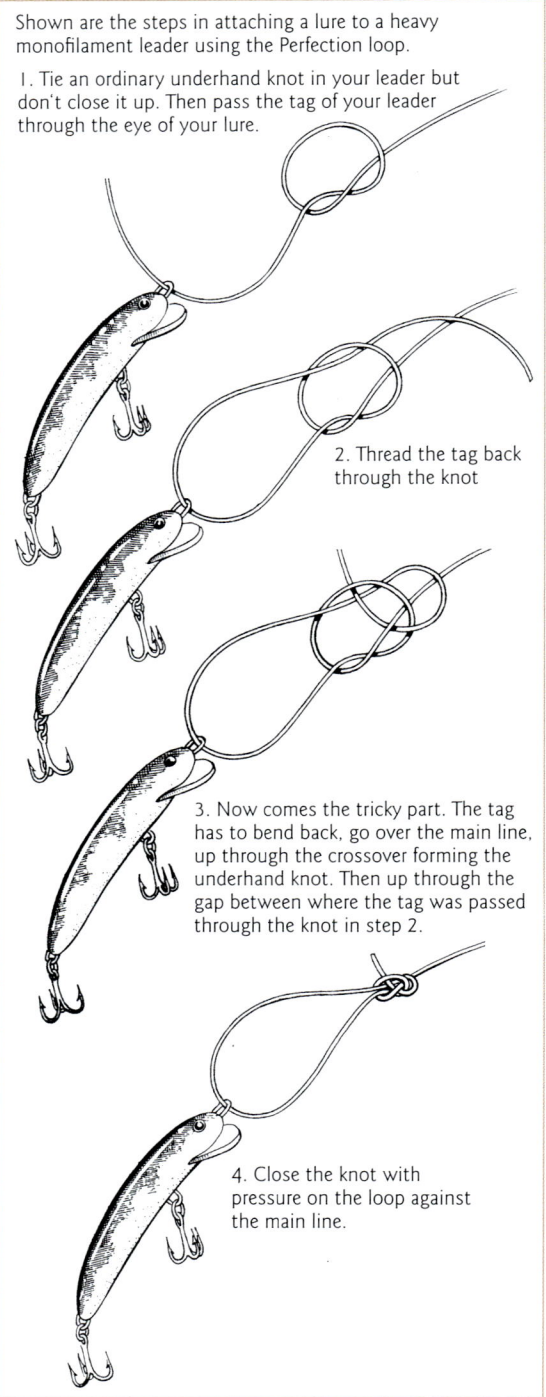

Shown are the steps in attaching a lure to a heavy monofilament leader using the Perfection loop.

1. Tie an ordinary underhand knot in your leader but don't close it up. Then pass the tag of your leader through the eye of your lure.

2. Thread the tag back through the knot

3. Now comes the tricky part. The tag has to bend back, go over the main line, up through the crossover forming the underhand knot. Then up through the gap between where the tag was passed through the knot in step 2.

4. Close the knot with pressure on the loop against the main line.

Figure 3.6: Rigging Minnow and bibbed lures for trolling. A loop knot is generally a good choice if you want to get the best action from your lures. Because the knot does not lock down on the lure's tow eye it gives the ability for lures to swing freely. Another option for attaching lures to your trolling line is to use a small (size 12 or 14) Duo Lock snap or one of the imported Norman Speed Clips.

Lures for Trolling 27

Spoons for trolling are relatively inexpensive and come in a huge range of shapes, sizes, colours and weights.

running a ball bearing swivel with most spoons as it may allow the spoon to spin, which means it won't catch fish. As with any trolled lure, always check the lure's action in the water at the speed you want to troll to make sure you don't get line twist and to get the best possible action!

No discussion of spoons or trolling for that matter would be complete without a mention of our homegrown Cobra or 'Tassie' lures. This style of unique spoon can also benefit from a little tweaking. Like most other spoons, by increasing the bend you can make these lures work better at slower speeds. Again check the lure at your desired trolling speed at the side of the boat. To make it easier to bend the lure, put it in a cup of warm (not boiling) water for a minute and you'll find it makes the job much easier. Don't overdo it, too much of a bend will cause the lure to spin out of control, resulting in serious line twist.

of pliers to produce an S-shape in the lure. Most spoons are fairly flat out of the box and can do with a bit of bending. Generally, the more of an S-bend you produce in the spoon, the better its ability to handle slow trolling speeds. Flat spoons will handle very fast trolling speeds (3 km/h or more) easily, but are not much good for very slow trolling. The most effective speeds for big fish will usually run from 1 to 2.5 km/h and will require more bend. This style of lure will catch fish from flatlines, planer boards, lead core lines and downriggers. If you run a spoon on a flatline it often helps to run a little weight (from 4 to 10 g) about 1 m up the line from your presentation to give the lure a little added depth. I would avoid

Often overlooked are the 7 g offerings from Wigston's and Lofty's Cobras. These smaller lures are great for targeting shallow water fish or over the top of shallow weed beds; in addition they are often a better imitation from a size standpoint than larger lures. I also prefer to run single Siwash hooks on this style of lure.

SMALL LURES

Small lures such as the Rebel Crickhoppers or Hellgromites, and small Flatfish lures that come

Small lures are often a good alternative presentation when fish are feeding on small baitfish or insects.

with two tiny trebles can be improved by changing hooks. Remove the two trebles (size 10) and replace with one size 6 treble or a size 4 Siwash hook on the rear of the lure. This doesn't seem to alter the lures' action or how they ride in the water, but it makes them far more productive when hooking big fish. Another option for small minnow lures like the 3 cm Rapala is to use a single hook as explained above (attached to the rear of the lure) or use a long shank hook attached to the front eyelet. To rig this type of set up, use an extra long shank hook and attach it to the front eyelet with a split ring or directly to the eyelet. To do this you will have to use a pair of pliers to open up the eye of the hook. Use a small rubber band to hold the hook up tight against the lure body. Depending on the length of the particular lure body and hook you can either run the hook point up or down. Rigging a lure in this fashion can give you a presentation that is far less prone to getting snagged.

Targeting big fish can often be frustrating. You can spend a lot of hours trolling with few fish to show for your efforts. To be really successful will mean honing all your skills and being prepared to experiment. It may seem like a lot of bother, but paying attention to the detail may just help you bag that trophy.

Successful flatline trolling is a result that comes from perseverance, knowledge and the ability to adapt to changing conditions. If you try to run a specific lure at a specific depth at a specific speed all the time your results will suffer. Keeping an open mind when trying new techniques and gear as well as a bit of trial and error with your presentations can result in more fish in the boat.

ALTERNATIVE RIGS FOR TROUT AND SALMON TROLLING

Trolling Flies

Trolling flies are a good alternative trolling presentation for many applications and conditions. Trolling flies, due to their inherent lack of weight or a bib, need to be used in conjunction with an attractor, trolling sinker or snap weights. One of my favourite choices of presentation for clear low water conditions is to troll a fly behind some type of an attractor, usually a dodger or flasher. Flies for trolling applications are designed to look their best when moving underwater. Most dry flies struggle to provide a good imitation of a bug or insect when submerged. Trolling flies should provide the right mix of colour, either bold or subdued, the suggestion through shape of the prey your target species is actually feeding on, and movement to attract fish.

The tying of trolling flies can include almost any type of material. Tying materials like chenille work well because they will soak up water and make the fly heavier. Adding material like tinsel or Mylar for contrast can also be useful. Heavier flies are useful because they help keep the fly from trying to float toward the surface. Tube flies are a good example of this phenomenon. Keeping the trolled fly on the same plane as the leader pulling it, is what a good presentation is about.

The choice of leader material between the dodger and fly for this technique is very important, and can spell the difference between fish in the

Trolling flies such as the large tube flies depicted are an excellent choice for trolling behind a dodger or flasher.

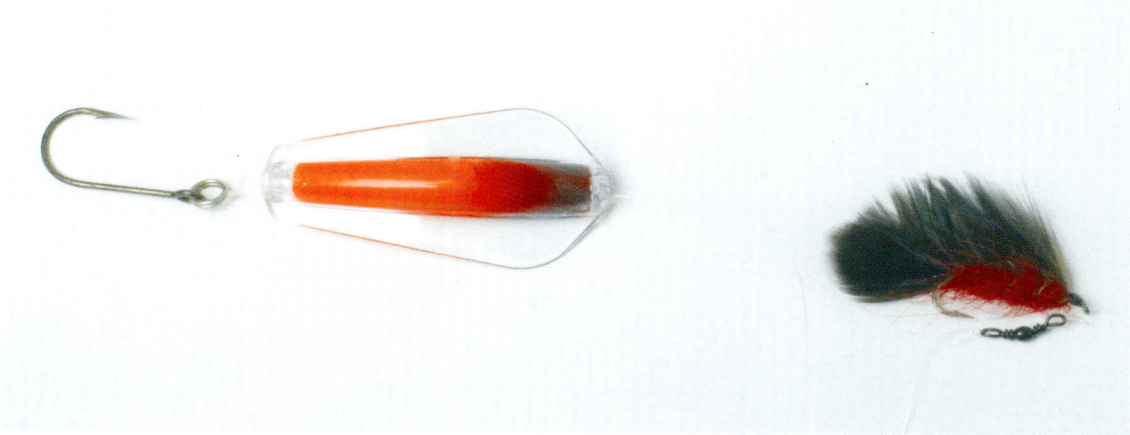

Figure 3.7: New Zealand method of rigging a fly above a trolled Cobra style lure.

boat and going home fishless. Most flies (unlike bibed lures) have little or no action when trolled on their own and your choice of leader determines what action from your attractor is imparted to the fly. As a generalisation long leaders in the 0.4 m–1.0 m range are too long for use with flies. Short leaders between 0.2–0.3 m are usually your best choice. This is one application where the choice of a stiffer line is the preferred option, as stiff short leaders will impart more action to the fly than a long limp leader.

A technique commonly used in New Zealand when trolling Cobra style lures employs a fly to cover swivels rigged above the lure. Rigging this style of lure with a swivel 500–800 mm above the hook allows the lure to slide down the line when a fish is hooked. This helps keep a hooked fish from being able to use the weight of the lure to dislodge the hook. Although not used as much here in Australia, this is a very effective method, with the fly often taken by a fish in preference to the lure.

Trolling flies can be presented on flatlines, with the aid of snap weights, lead core lines or downriggers to fish a range of depths and conditions. Trolled behind a dodger, flasher, or bladed attractor they can be extremely effective. I have successfully used flies up to about 100 mm long to target both trout and salmon. With the type of clear low water conditions faced during autumn, long dropbacks are usually the go to avoid spooking your targeted fish. The drawing below depicts a typical set up for rigging flies and attractors.

Alternative Rigs

One of the great advantages of trolling is the ability to cover a large amount of water when searching for fish. To really take advantage of this phenomenon it pays to include as many trolling and rigging techniques in your bag of tricks as possible. The old

Rigged behind an attractor, trolled flies can be productive for many species under a range of conditions.

adage about necessity being the mother of invention really comes into play when you're out on the water scratching your head trying to figure out what to try next. I certainly can't lay claim to having developed the following methods of rigging, but I can attest to their effectiveness at catching fish!

To effectively troll the margins of our deeper lakes and some of the shallower (2–8 m) lakes most anglers opt for flatline trolling shallow and deep diving lures to target both trout and our native species.

Different seasons, water conditions and levels as well as water clarity should dictate how you approach the selection of lure or bait and presentation. During low water or summer conditions a large part of many fish species' diet is made up of aquatic invertebrates. Dragging around a big deep diving lure is not likely to be all that effective if the fish you are targeting are feeding on small invertebrates on the bottom.

An alternative to trolling with lead core line or using a downrigger (though this technique can be used effectively with downriggers to avoid hanging up your bomb on snag-infested waters) this rigging method was developed as a back-trolling technique on North American rivers. It works exceptionally well in still water and allows you to target fish in the 2–8 m range.

The heart of this rig is a deep diving lure, not to catch the fish but to put your presentation in the strike zone. This is an ideal time to drag out those old cod lures that have been relegated to the bottom of your tackle box. Our locally made deep divers are ideal for this purpose. They dig in to the bottom and keep your trailing lure or bait where it needs to be, right near the bottom. You also have the advantage of being able to use the rod sweeping technique for clearing snagged lures that you employ when trolling for Murray cod or other native fish.

Rigging

To set up this rig for trolling for trout I generally use 4–5 kg monofilament as a main line, and 6–8 kg braids if I am chasing natives. Choice of a main line should be determined by the bottom condition of the water you're fishing. Braided gel spun lines will give you excellent feel for what's happening at the business end of the rig, but if you are fishing very clear water then mono may be a better choice. On the end of your main line tie a snap swivel to connect your diving lure. The diving planer you choose to employ for the depth of water you're fishing and the water clarity should govern this technique. Diving planers such as Fish Seekers and Luhr Jensen's Jet Divers, can also be employed for this task but I prefer to use lures to enable me to use lighter tackle. Make sure you remove the hooks from your diving lure, it will help keep it from getting hung up and its main use is to get your presentation down to the fish. In most cases I use lures that I've painted a fairly neutral or drab colour for this purpose. StumpJumpers, Boomerangs and Oar Gees all work well for this purpose but I've also used Mudbugs and Storm Hot N Tots. In this instance I want the diving lure to get my trailing lure fly or bait down near the bottom, not to act as an attractor.

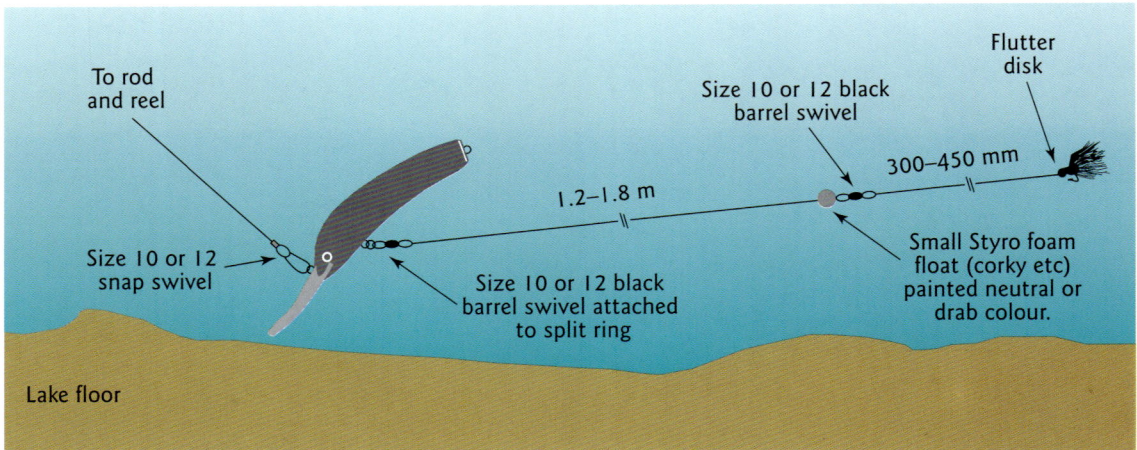

Figure 3.8: This rig utilizes a deep diving lure to get your presentation down near the bottom. This technique is very effective when trolled in either impoundments or rivers.

Lures for Trolling

Flutter discs are small round (10–25 mm diameter) plastic discs that can add considerable action to trolled flies. To make your own, simply find some thin plastic sheet (blister packs from lures and other similarly packaged merchandise works well) anything up to about 1 mm thick works well. Mark a diameter and centre using a compass and cut out with a pair of scissors. Drill or punch a hole in the centre of the disc to thread your line through. A 1 mm hole is usually enough. Once you have cut out the outside diameter of the disc, pinch or fold the disc in half to create a crease down the middle. This crease causes the disc to have a slightly concaved surface one side. When rigging, place the concave side in the direction you will be trolling your bait or lure.

My choice of trolling flies for this technique generally includes any imitation that resembles what a bottom feeding fish will be looking for: shrimp, nymphs or any crustacean patterns can be excellent choices in clear, low water levels. I will sometimes run streamers or minnow patterns in brighter colours as an alternative during rising water levels or discoloured water after heavy rain.

One drawback with trolling flies has been their lack of action unless trolled behind an attractor to give them some life. To overcome this problem I employ what used to be called flutter discs. Known also as action discs flutter discs are available commercially, but making your own is simple and costs next to nothing. The picture explains how to construct them, and it is well worth the little effort required to make them, as they really make a trolled fly come to life!

The following figure (Figure 3.9), depicts another rigging method that utilises trolled bait and will work in almost any type of conditions. This rig is one of my favourites for rising water levels or when a lake or river has become muddy or discoloured. Scrub worms or yabbies are ideal for this rig, but being heavier than trolling flies they need a little different approach for rigging. With scrub worms I sometimes use a hypodermic needle to inject air into the worm for added floatation. This is a technique bait anglers use for floating a worm above a weed bed or structure. To rig scrub worms or yabbies I usually put three or four plastic beads in front of the bait. I also like to use Spin-n-Glows in front of a bait to add a little extra attraction when the water is discoloured. Spin-n-glows are a foam body attractor with Mylar wings that rotate and provide floatation and attraction to trailing baits or lures. You can also use flutter discs in front of your bait to give them a little extra action.

When using either of these rigs I often hold the rod to get a greater feeling for the contours of the bottom and to be able to react if the lure hangs up on a snag.

This technique is very much like trolling for Murray cod, the big advantage is you can target several different species. Keep a very close eye on your depth sounder. Even in shallow water it can give you a few seconds advance warning as to changes in the bottom contours, structure and potential snags. I use one of the new generation Lowrance LCX 16 sounders and it is absolutely amazing. The clarity and definition are so good you almost think your watching television. It makes watching bottom conditions a breeze! When you first set this rig up it may seem a bit fiddly, but it soon becomes quite easy to rig and use. Have a go, this technique can produce big bottom hugging fish!

Favourite Freshwater Lures

FISH SPECIES	FLATLINE TROLLING	DOWNRIGGING	DEEP DIVING LURES
BROWN TROUT	1. Rapala Originals 2. Rapala Husky Jerks 3. Storm Thunderstick Jr. 4. Tassie lures 5. Various Spoons 6. Predatek Min Min	1. Rapala Originals 2. Jointed Rapalas 3. Tassie Lures 4. Legend Helmax 5. Dodger/Flies 6. Flatfish/Quickfish	1. Rapala Deep Husky Jerk 2. Storm Thunderstick Jr. Deep 3. Tassie Lures 4. Legend Helmax 5. Australian Crafted 6. McGrath
RAINBOW TROUT	1. Rapala Original 2. Rapala Tail Dancer 3. Various Spoons 4. Tassie lures 5. Rapala Husky Jerk	1. Rapala Originals 2. Jointed Rapalas 3. Flutter Spoons 4. Streamer Flies/Dodger 5. Tassie Lures 6. Flatfish/Quickfish	1. Rapala Deep Husky Jerk 2. Tassie Lures 3. Legend Helmax 4. Australian Crafted 5. McGrath
GOODOO/ MURRAY COD	1. Oar-Gee Plow 2. StumpJumper 3. Australian Crafted 4. Predatek Boomerang 5. Storm Hot'N Tot 6. Luhr Jensen Hot Lips 7. Spinnerbaits 8. Soft Plastics	1. Oar-Gee Plow 2. StumpJumper 3. Australian Crafted 4. Predatek Boomerang 5. Storm Hot'N Tot 6. Spinnerbaits	1. Oar-Gee Plow 2. Australian Crafted 3. StumpJumper 4. Predatek Boomerang 5. Storm Hot'N Tot 6. Luhr Jensen Hot Lips
GOLDEN PERCH	1. Storm Hot'N Tot 2. Oar-Gee 3. Australian Crafted 4. StumpJumper 5. Merlin	1. Oar-Gee 2. Australian Crafted 3. StumpJumper 4. Merlin 5. Storm Hot'N Tot	1. Rapala Shad Rap RS 2. Legend 3. Oar-Gee 4. Australian Crafted 5. Merlin
AUSTRALIAN BASS	1. Oar-Gee 2. Australian Crafted 3. Legend 4. Merlin 5. Rapala Fat Rap	1. Oar-Gee 2. Australian Crafted 3. Legend 4. Merlin 5. McGrath	1. Australian Crafted 2. Mann's 5+ 3. Rebel Crawdad 4. Spinnerbaits 5. Predatek Min-Min
CHINOOK SALMON	1. Tassie Lures 2. Rapala Original 3. J Plugs 4. Flutter Spoons 5. Rapala Countdown	1. Flutter spoons 2. Rapala Originals 3. Jointed Rapalas 4. Streamer Flies/Dodger 5. Tassie Lures	1. Rapala Deep Husky Jerk 2. Tassie Lures 3. Legend Helmax 4. Australian Crafted
ATLANTIC SALMON	1. Streamer Flies 2. Flutter Spoons 3. Tassie Lures 4. Rapala Original 5. Rapala Jointed	1. Flutter Spoons 2. Streamer Flies/Dodger 3. Tassie Lures 4. Rapala Originals 5. Jointed Rapala	1. Rapala Deep Husky Jerk 2. Storm Thunderstick Jr. Deep 3. Tassie Lures 4. Legend Helmax 5. Australian Crafted
REDFIN	1. Rebel Crawdad small 2. Celtas 3. Spinning Blades 4. Small spoons 5. Rapala Originals 6. Soft Plastics	1. Rebel Crawdad small 2. Celtas 3. Spinning Blades 4. Tassie Lures 5. Rapala Originals 6. Soft Plastics	1. Rebel Crawdad Deep (small) 2. Celtas 3. Spinning Blades 4. Tassie Lures 5. Rapala Originals 6. Soft Plastics

Note: Tassie Lures refers to Lofty's Cobras, Wigstons Tasmanian Devil, Tillans King Cobra, Sting Lures and Johnson Lures in sizes from 7–25 grams.

Figure 3.9: Bait Trolling Rigs

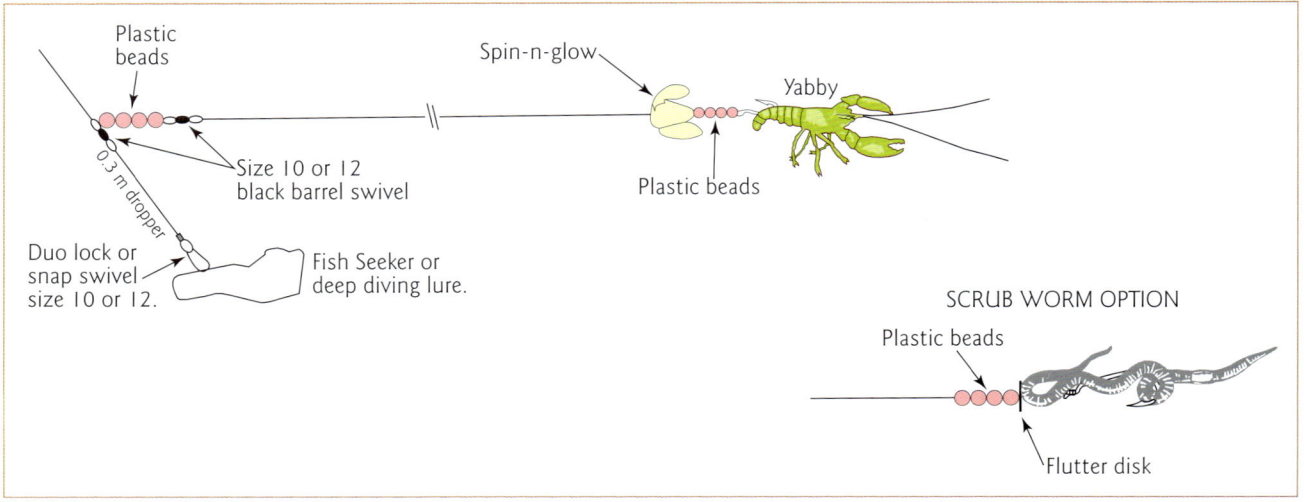

CHAPTER 4
Surface Trolling
FLATLINING

Surface or flatline trolling can be effective from virtually any size or style of boat. Boats fitted out with as much gear as the example in this photo can accommodate almost any type of freshwater trolling.

FLATLINE TROLLING FOR TROUT AND SALMON

Trolling is without doubt one of the most popular effective and challenging methods of angling that exists today. Trolling for trout with light tackle can be as sporting and enjoyable as fishing by any other method. Flatline trolling (without the use of anything but a rod, reel, line and lure or bait) can be a real buzz when you're hooked into a good fish. If you want to improve your catch rate, there is a wide range of issues to look at to improve your trolling techniques.

There are a number of variables that have

Careful handling of fish to be released will ensure that other anglers will have the opportunity to catch our species in the coming years.

a direct effect on lure depth when trolling, and fortunately with the exception of lure design, we have control over these variables. How well we understand all the variables and how well we are able to apply them to our trolling has an enormous impact on our success. Looking at these variables, including how you position your boat, lure design, line diameters, line out and trolling speed we should be able to develop a better understanding of what happens on the business end of our line.

RODS AND REELS FOR FLATLINE TROLLING

Almost any style of rod and reel can be used successfully for trout trolling, but a bit of forethought about your set-up can help eliminate potential problems. A typical set-up for a small boat with two anglers aboard would usually include two rods each (regulations permitting). In this scenario the norm for rod placement (and of course, rod holders) would be to run two rods straight out the back over the transom and the two remaining rods at 90º (straight out the sides). This set up works very well, but can be helped by a bit of thought and careful selection of the gear you choose. Longer rods on the outside lines will give your lures a wider spread and enable you to make sharper turns with less chance of tangled lines.

Both spinning and casting reel advocates have their favourites for flatline trolling, and both obviously have advantages and disadvantages. I generally opt for casting or overhead reels to run out the back as flatlines. Overheads trolled in this manner tend to sit well in a rod holder with the guides facing up and any striking fish pulling line from the reel against the backbone of the rod as it should be. Rods for this application should be in the 1.8–2.1 m length range or whatever you feel comfortable with. Fast action, light rods in the 2–4 kg range work well for this application. Flatlines run at 90º to the gunnels are, I believe, another set up or scenario that needs careful consideration. Short light rods are great fun to land fish with, but they don't really give your lures much of a spread. To increase the distance between my trolled lures I generally use a much longer rod that allows me to turn more sharply and work my lures to better advantage. Light fast action overhead rods in the 2–4 kg range with a length of 2.2–2.4 m are ideal for this application, but are almost impossible to find

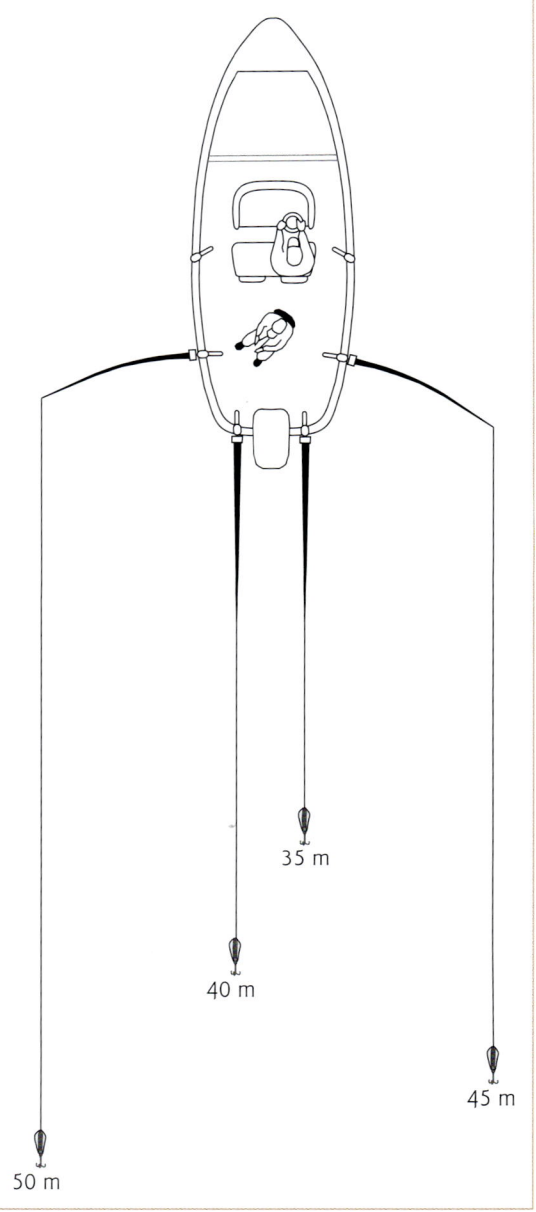

Figure 4.1: Running 4 trolling lines from a small boat is relatively easy with the right gear and a little preparation.

in Australia. I've found the steelhead rods made for the North American market suit this technique perfectly, but they are not always readily available. Most major rod producers including Loomis, Daiwa, Shimano, Quantam, Penn, Reddington and Shakespeare all make these rods in both overhead and spinning configurations. Ask your local tackle shop to see what they can get.

Whether flatlining or downrigger trolling rods need to have a soft enough tip to allow the angler to be able to see the action of a trolled lure.

LINE DIAMETERS

Choosing a trolling line for most anglers generally involves looking at properties like strength, abrasion resistance, and amount of stretch and, for some applications, colour. There are a host of excellent trolling lines available, choosing one is partly personal preference and partly the application. To determine the difference line diameter makes in relation to lure diving depth, a general fixed standard has been set at 10lb breaking strain. Studies done to establish diving depth of lures have used this standard as a baseline to determine accurate depths. Nearly all lures will run deeper when trolled on smaller diameter lines, due largely to the effect of water resistance against the line. All 10lb test lines are not the same diameter; they can vary in diameter considerably. I've had 10lb lines with diameters of 0.20 mm through to lines with a diameter of 0.35 mm still rated at 10lb breaking strain! Virtually all my flatlining for trout is done using monofilament with a diameter of 0.20–0.22 millimetres.

The new generation of braided (gelspun)

Successful trolling with lures requires a lot of thought, planning and work. How you use your equipment, especially your lures, will determine your level of success!

multi-filaments are also another viable option for flatline trolling. The latest research that I've seen seems to indicate that there is little difference between 6 and 10lb braid in diameter and trolling depth performance. Either breaking strain will add approximately 25% to the depth of your trolled presentation. Softer rods and a light drag are the order of the day for flatlining with braids. The minimal stretch with braided lines can result in hooks pulling and dropped fish when you are not familiar with this style of line. The extra depth that can be achieved on a trolled lure utilising braids can give the trolling angler the option of prospecting more of the water column.

The new fluorocarbon lines are a bit expensive to run as a main line, but I've found them to be great as leader material. With light reflecting qualities similar to water it is claimed that this material is virtually invisible in water. I've used the Rio Fluro-Flex as leaders and it has proven to be tough and very durable.

LINES

One of the most critical factors in achieving optimum performance from any lure is line selection. The line you choose for a particular application deserves a lot of careful consideration, after all it's your main connection to a fish. Many anglers still seem to pay little attention to the detail when it comes time to spool up for a fishing trip. The age-old adage about getting what you pay for certainly rings true when it comes to line selection.

The technology and choice available to anglers in the form of monofilaments, co-polymers, braided gelspun, and fluoro carbon lines is really quite staggering. The array of manufacturers and brands of lines, along with the price and availability, seems to have expanded enormously in the past ten years. The most important areas for consideration with trolling applications should be line diameter, stretch and abrasion resistance.

When you are trying to decide on a line for trolling, keep in mind that lines that are primarily made for lure casting, are not the answer for trolling. In many of our impoundments you will encounter submerged trees, rocks, fences etc. that will really test out your line. Trolling lines need to be fairly tough with good abrasion resistance and low stretch, not soft and limp like a good casting line. The brand you choose is a personal preference, but buy as good a quality line as you can.

Generally line diameter for downrigging is not as crucial as when surface trolling. Having said this you still need to choose your line carefully. Your selection of line for downrigging should take into

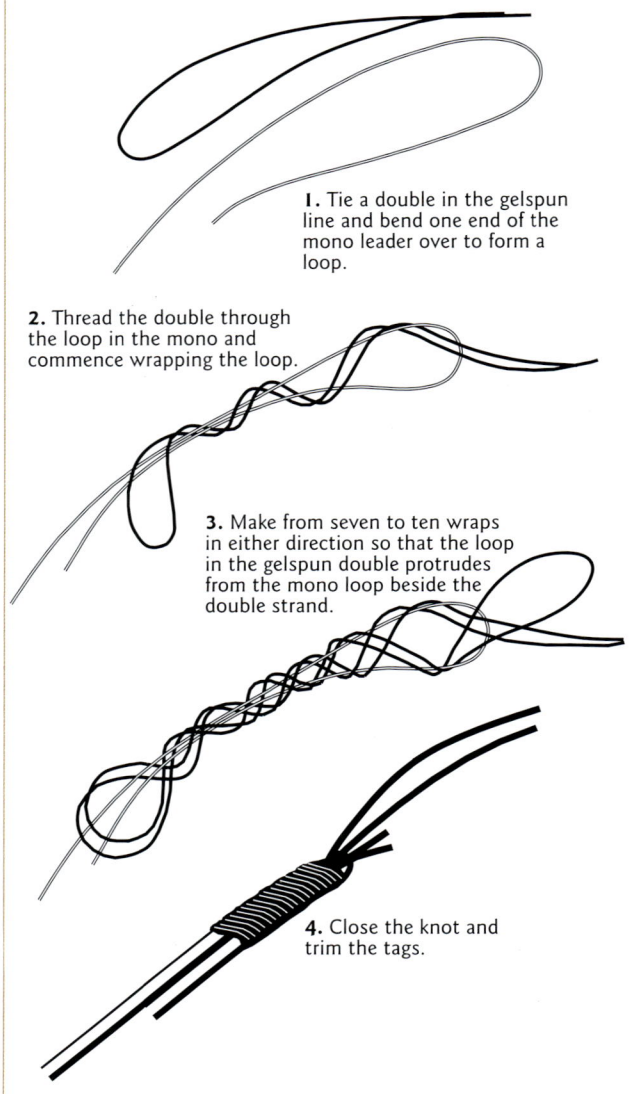

Figure 4.2: The drawing depicts an excellent method for joining braided gelspun line to a monofilament or fluorocarbon leader.

consideration what type of line release clip you use and the type of downrigging you intend to do. The belly in your line (the main line from the rod and reel to the release clip) is created by water pressure pushing against the line as you move through the water. The amount of belly in the line increases with larger diameter lines, so it pays to not go overboard with heavy lines for this application. If you are constantly chasing the bottom in snag ridden waters you need to consider a tougher abrasion resistant line. For most Australian conditions line from 0.20

to 0.30 mm diameter will handle just about any fishing situation for trout or salmon.

Line diameter plays a major role in how deep your lure runs when trolled or retrieved. Thinner diameter lines allow your lure to run deeper and optimise the lure's built-in action. In most applications diameters in the 0.18 to 0.22 mm range are your best bet for trolling most trout style lures on surface or flatlines. My personal choice for flatlining for tout is line of 0.20 mm diameter. Lines of this size (and of course all others) can vary enormously in breaking strain from one maker to another. Most major brands are between 2–4 kilograms breaking strain for 0.20 mm line, which is more than adequate for most trout trolling applications in our impoundments.

Today's new generation of gelspun braided lines are amazing to use. Surface trolling or flat lining with braided lines is almost like learning to troll again. The amount of stretch in this type of line is so minimal as to be almost non-existent. Your first hook-up on a fish will amaze you with what you can feel; every headshake or flap of a fish's tail is transmitted through this type of line. If you have never used this type of line keep in mind that with so little stretch you need to run a very light drag and use a rod that is soft enough to be forgiving and can act like a shock absorber. Lines such as Fireline, Spiderwire, Fins and the Australian-made Platypus are all excellent performers.

Line stretch and resultant breakage can be a problem with some lines. Stretch is usually difficult to detect until you snag up or try to land a big fish. Ordinarily you can see or feel the difference when your line has stretched and not fully recovered. When your line is stretched it may also show up as a different colour, most noticeable with darker coloured lines. To avoid problems like this most line manufacturers recommend that you cut about 1.5–2.0 m off your line and retie your lure after landing a big fish or getting snagged. It's probably something that most anglers don't do often enough.

Surface or flatline trolling with lures can be very productive if used wisely.

Trolling Speed

In years past there has been an enormous amount of information written about trolling speed's effect on lure running depths. I've read opinions that varied from speed having no effect on lure depth, to it being absolutely critical. If we think about the fairly limited speed that we employ for freshwater flatline trolling (usually from 1–5 km/h) I would suggest that the real answer lies somewhere between these two extremes for virtually all floating/diving lures.

The two most common methods of measuring trolling speed are the surface paddle wheels on depth sounders with a speed/temp indicator and the 'speed over ground' (SOG) indicator on most GPS units. Neither of these indicators is infallible and should be used as a guide or indication only. When you have both boat movement and the water movement (wind or wave action) this invariably affects paddle wheel speed indicators, making it difficult to get an accurate reading of speed. Modern GPS units should provide accurate speed information however due to the 'selective availability' of GPS, a government security manipulation of the GPS signal, day-to-day fluctuations occur in GPS signal accuracy.

Given the inaccuracies from most mechanical/electronic indicators, how do we get a reliable indication of speed? At the end of the day, the most reliable estimate of speed appears to come from a combination of instrument measurements and experience. I've little doubt that trolling speed is important, but also believe that, how important will depend on your particular application. Speed is critical in controlling the depth of negatively buoyant (sinking) lures, but not as critical in determining the running depth of floating/diving lures trolled within the range of speeds we normally use. If we think of our speed in broader terms such as slow (up to 1.5 km/h) medium (1.5–3.0 km/h) and fast (3–6 km/h) we have the opportunity to apply this range of speeds to our trolling. Slow (up to 1.5 km/h) speeds are often most productive in cold water conditions or when trolling live bait, such as mudeyes or scrub worms. Medium speeds (1.5 – 3.0 km/h) are probably the most used speed for trolling a vast array of lures. Fast trolling speeds (3-6 km/h) are often best applied when fish are very active or used as a triggering technique to tempt a strike from reluctant or inactive fish. Again these speeds should be used as a guide only; there will be exceptions to every rule.

Line Out (dropback)

The single most important factor in controlling depth of a flatline trolled lure is the distance we let the lure out back behind the boat. This line out or dropback is the easiest variable for us to control. Letting more line out while moving, all lures will eventually reach an effective maximum running depth. The majority of floating/diving lures designed for trout and salmon will reach their maximum depth at 50–75 m of line out. Obviously, if we know the distance a lure is behind the boat and the relationship to the distance the lure is down, we can achieve any depth in between the lure's maximum depth and the surface. To do this accurately we need some sort of mechanism to measure line out. There are a number of means to do this including marking your line in measured

Line out or dropback is a major factor in determining how deep your lure will actually dive on a trolled flatline.

increments, counting bars on a level wind reel, line counter reels, or a Tackle Tracka line counter. Regardless of which method you use, the accuracy and consistency of measurement will determine how accurately you can predict the depth of your trolled lure!

SURFACE TROLLING FOR NATIVES

Trolling for Australian native freshwater species is not a pastime for the faint hearted! Fish such as golden perch and Murray cod are by their very nature ambush predators. They spend much of their time in and about structure including submerged trees, rocks, shelves and drop offs waiting for a potential meal to wander by. To effectively target these fish while trolling means you will constantly be putting your lure or bait at risk of being snagged up or broken off.

The last ten years has seen an enormous growth in anglers chasing our native fish especially Murray cod and golden perch. Trolling for these fish is a technique that shares many common details with methods of trolling for other species, with some significant differences.

TACKLE FOR NATIVE FISH TROLLING

Trolling for natives requires serious tackle compared to fishing for other freshwater species. Whether you fish lakes or rivers the tackle necessary needs to be robust. When you think about the possibility of encountering fish up to 40 kg (close to 100lb) the need for this kind of tackle becomes obvious.

RODS FOR NATIVE TROLLING

Rods rated to 8–10 kg (17–25lb) and in lengths from 1.8–2.1 m are a good starting point for trolling for natives. Rods with plenty of backbone (for muscling big fish out of the snags) and a fairly light tip are vital for this technique. A light tip allows the angler to visually check if the lure is working properly. Tip action can alert the observant angler to any weed or rubbish fouling your lure, and will also enable you to see the action of smaller lures. If you want to increase your chances of landing that once in a lifetime fish, steer away from the temptation to use a rod that is great for casting to troll with. Both techniques make different demands on a fishing rod; the right tool for the job might mean the

Huge fish such as this native Murray cod are the dream of most anglers trolling for our native fish.

difference between landing and losing that trophy fish! Most major rod producers including G.Loomis, Daiwa, Shimano, Quantam, Penn, Reddington and Shakespeare all make these rods in both overhead and spinning configurations.

Reels for Native Trolling

For trolling applications I firmly believe that a sturdy robust baitcast or overhead reel should be your first preference if you are looking to set up a reel for trolling for Murray cod. As mentioned previously, tackle for these fish needs to be fairly heavy. I would personally opt for a strong reliable round overhead reel for trolling as opposed to a low profile casting reels. This may seem like a bit of overkill but if you are specifically targeting big Murray cod you need the right gear. With few exceptions the small low profile casting reels that are now available in Australia are primarily designed for casting and retrieving lures to large mouth bass in the US market. Most of the bass targeted in the US run from 2–10lb (1–5 kg) with occasional larger fish thrown in, but you may be asking a lot to repeatedly expect these reels to cope with fish that can exceed 40 kilograms. Low profile reels can however, often be appropriate for trolling for other native species such as Australian bass or golden perch. Daiwa, Shimano, Quantam, Penn, Okuma and Shakespeare all make these reels in many different sizes and price ranges. Due to the popularity of this style of reel overseas a lot of companies have developed some excellent models that also carry some very hefty price tags!

Conventional overhead reels that will also do the job well (some of these reels don't have line counters or clickers) and that are a suitable size include the Abu-Garcia 5600C4 or 6500C4, Daiwa Millionaire CVZ or CVX 253 or 300, Okuma Convector, Shimano Calcutta and the Cabo and Classic baitcasters from Quantum.

Line

The development of gelspun braided lines has been an enormous benefit for trolling for Australian native fish. As most trolling for Murray cod involves bouncing lures off the bottom or structure such as submerged trees, braided lines give an immediate indication when you are hooked into a fish or a snag. When monofilament line is used the resulting stretch can make it difficult to avoid many snags. The thinner diameters afforded by using braided lines means better utilization of a lure's diving ability to achieve a given depth. A critical factor when trolling lures in heavy structure is line control, using braided line also enables an angler to run shorter drop backs to achieve a given depth and to manoeuvre a lure through the structure.

Leaders for Braided Lines

The major drawback with braided lines is their ability to cut into soft material like submerged partially decomposed timber and their fine diameter generally means rocks can cut them quite easily. To avoid this problem use a monofilament leader about a metre long tied directly to the braid. Fluorocarbon lines work very well for this technique and though somewhat expensive are well worth the money.

Terminal Tackle

There has always been a lot of debate about using a loop knot or a snap swivel to connect lures to your leader or main line to facilitate fast lure changes. This is one technique where the use of a snap is I believe appropriate. If you choose to use a snap there are a number of different styles to choose from. I have used Duo-Lock snaps with ball bearing swivels, Coast-lock snaps with ball bearing swivels, and Norman Speed Clips all with good success. Whatever type of snap you choose ensure it is the best quality you can buy. Cheap snaps opening up or straightening could cost you the fish of a lifetime. A good rule of thumb is to use a snap that is rated to 1.5–2 times the breaking strain of your line. This will ensure that your line will break before the snap.

Lure Retrieving

Lure retrievers are a useful addition to any troller's tackle box, but are absolutely essential if you troll for our native species. Trolling lures developed specifically for native fish can cost up to $20.00 each and can add substantial cost to your fishing trip if you lose a few in a day. There are several different types of retrievers available and all share some similarities. They are most effective when you can manoeuvre the boat to get directly over the snagged lure. This can sometimes be a tall order if you are fishing in very windy conditions or trolling in a river with strong current.

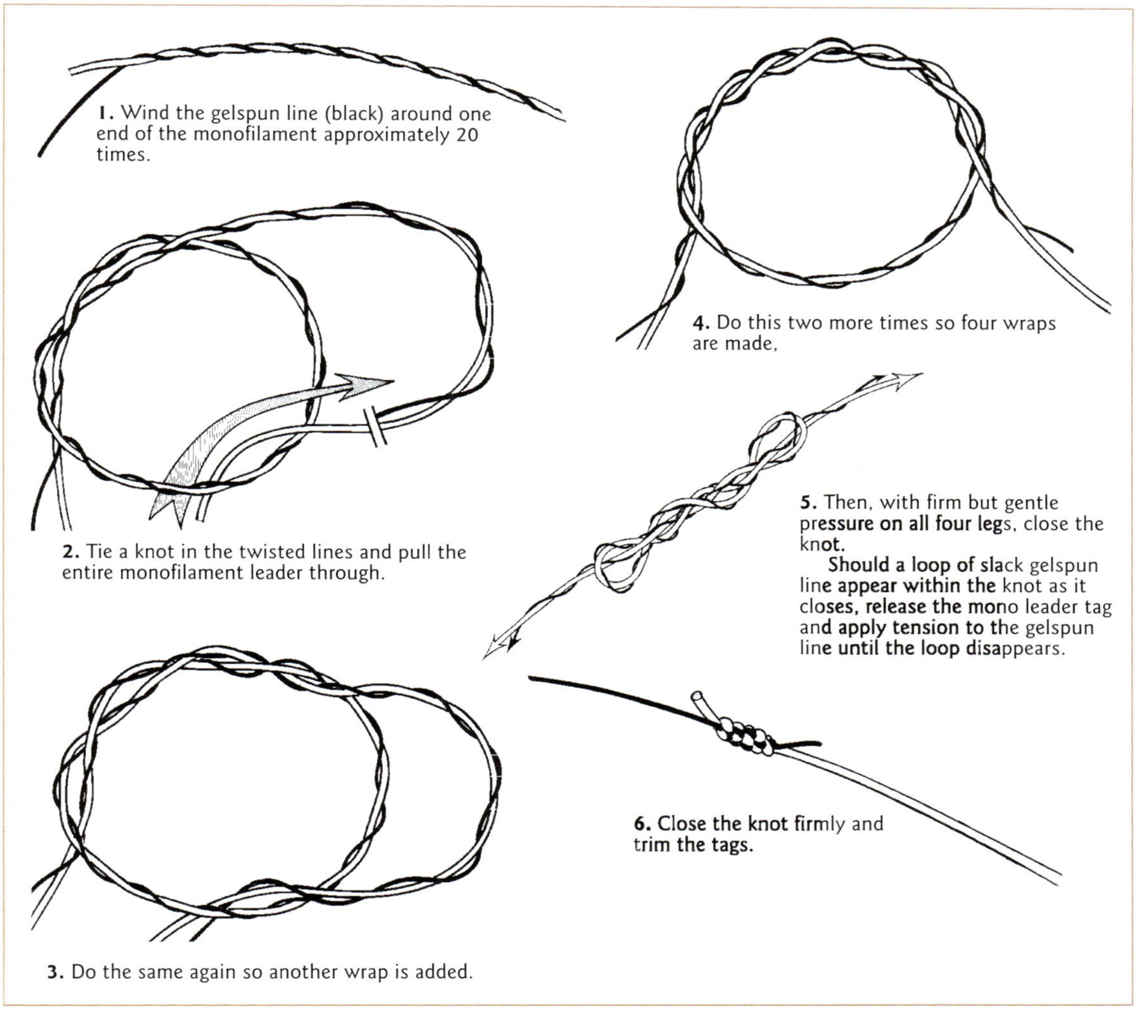

1. Wind the gelspun line (black) around one end of the monofilament approximately 20 times.

2. Tie a knot in the twisted lines and pull the entire monofilament leader through.

3. Do the same again so another wrap is added.

4. Do this two more times so four wraps are made.

5. Then, with firm but gentle pressure on all four legs, close the knot.
   Should a loop of slack gelspun line appear within the knot as it closes, release the mono leader tag and apply tension to the gelspun line until the loop disappears.

6. Close the knot firmly and trim the tags.

Figure 4.3 ABOVE: Tying monofilament to braided gelspun lines can be very effective with the right type of knot. The drawing depicts another option for making this connection.

RIGHT: Specimens like the two Murray cod pictured are the ultimate reward for anglers targeting our native species.
PHOTO: RICK HUCKSTEPP

Figure 4.4

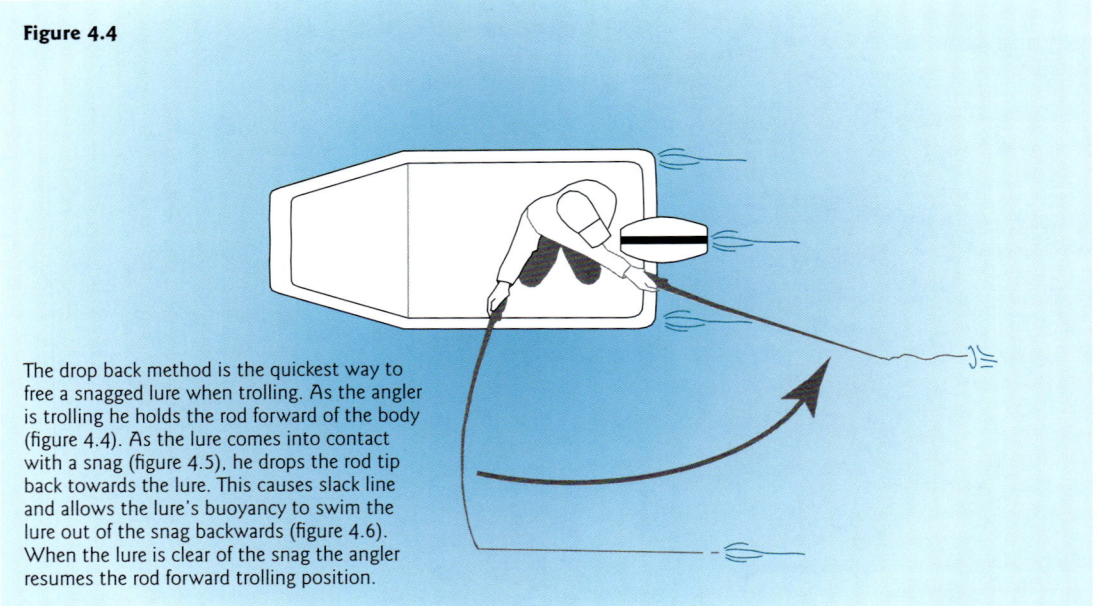

The drop back method is the quickest way to free a snagged lure when trolling. As the angler is trolling he holds the rod forward of the body (figure 4.4). As the lure comes into contact with a snag (figure 4.5), he drops the rod tip back towards the lure. This causes slack line and allows the lure's buoyancy to swim the lure out of the snag backwards (figure 4.6). When the lure is clear of the snag the angler resumes the rod forward trolling position.

Figure 4.5

Figure 4.6

The figures above and left illustrate the drop back method for retrieving snagged lures when trolling.

LURES FOR NATIVE FISH

To attempt to categorise lures for trolling for Australian native fish is a difficult task. The explosion of impoundments stocked with native species has seen a similar rise in the number of anglers chasing these fish and the lure makers producing lures for this market. Lure design for native species appears to have been primarily the chase for attaining greater depth. I would suggest that this is primarily due to the slow take up of techniques developed overseas for getting to greater depths (such as downriggers, snap weights etc.) but also because the style of fishing is unique to Australia. Hand holding a single trolling rod with a lure bouncing off the snags and digging into the bottom is the most effective means for getting a big cod to leave its lair!

Most trolling tactics involve using larger lures that are floating divers. Until recently these larger lures were employed to get to the depths where the big fish hang out. Larger lures often represent a more substantial meal than a small lure to a hungry predator. Large lures also offer

ABOVE: Flatline trolling for natives usually involves heavier tackle to troll many of the big deep diving lures for Murray cod.

RIGHT: Smaller ultra deep diving lures are becoming the focus of several native lure manufacturers. These smaller lures can give anglers more choice when targeting fish that are suspended in deep water. Some of these lures are reputed to be able to troll at nearly 10 m deep when trolled on fine diameter braided line.

the advantage of displacing more water as well as creating more noise to attract fish. Big lures tend to float off snags more readily than their smaller counterparts, due to their greater size and resulting increase in positive buoyancy.

Many manufacturers of native lures are now producing ultra deep diving smaller lures for anglers targeting species like golden perch and Australian bass as well as Murray cod. These small lures can be very effective in the right conditions and provide a smaller profile that may be more appealing to species like golden perch and Australian bass.

Bladed Lures

Bladed lures have been around for a long time now. The best known of this style of lure is the Aeroplane Spinner, still produced today, and still catching fish!

Bladed lures for native fish have been and continue to be very effective lures for trolling.

Although primarily considered by many anglers as a casting lure, spinnerbaits are another productive bladed lure. Spinnerbaits are now available in a range of weights, sizes and blade configurations that trolled slowly can produce fish.

Trolling Speed

Speed for trolling native lures is largely dictated by your choice of lure. Very large lures for Murray cod can generate a lot of water resistance when trolled. If you try to troll these lures too fast they will generally blow out of control or pull so hard that you'll be struggling to hang on to your rod with both hands. The best indication of trolling a lure at the right speed is to lower it into the water and start trolling feeling the lure's action. The use of braided lines makes this fairly easy. As you increase the boat's speed you will be able to tell if the lure is working properly, digging in and tracking in a straight line. Speeds in the order of 0.5–1.5 km/h are a good place to start for trolling big lures.

CHAPTER 5
Planer Board Trolling

Open water trolling with planer boards can give the angler the option of trolling multiple lines well away from the boat's path of travel.

TROLLING WITH PLANER BOARDS

Planer boards have been around now for a number of years, though here in Australia they still don't have the following that they enjoy in North America. Trolling is not just a means of making presentations at almost any depth below the surface of a body of water with downriggers, weights, paravanes etc., but a very effective shallow water technique, primarily running flatlines or surface lures with the occasional deep divers. The use of planer boards seems to add another layer of complexity (or it seems to be perceived by a lot of anglers as being complex) that many trollers don't seem to want, so in turn they miss a lot of fish. Trolling boards are operated like a downrigger, except they run out horizontally (to the side of your boat) where you can see what is going on. Instead of lowering a weight into the water, a

Planer boards can be used on almost any size craft and will enable you to target fish in those hard to reach areas, such as shallow water shorelines.

board is placed on the water (they float) and the line is released to allow them to plane off to the side of the boat.

While downriggers allow the angler to fish any depth of water, trolling boards are primarily used to fish the top 5 m of water and allow lures or baits to be trolled well away from your boat (5–50 m to the side). This method has the distinct advantage of allowing the angler to run a lure or bait in undisturbed water, close to a bank, or in water too shallow for your boat to navigate.

When fish are using the top 5 m of the water and your boat comes into it you'll find that fish do just what you'd do if you were standing in the path of a slow moving car headed in your direction, you'd move to the side and get out of the way. Fish do the same thing, get out of the path of this oncoming monster (your boat).

Any angler who has fished for any species of fish by trolling knows that your line is going to trail in the path of the boat. Your lure is going through an area of water you just chased all the fish away from! For many years saltwater anglers have understood and taken advantage of this phenomenon by using outriggers to spread their lure or bait presentations.

To troll productively with flatlines relies on two main factors, how much line you have out behind the boat and how you work or manoeuvre the boat to position your lures or bait. The usual rule of thumb when flatlining in clear shallow water is to run long drop backs to avoid spooking fish. The more shallow the water and the greater the amount of boat activity, the more spooked fish can become, requiring even longer lines. Rarely can you motor through shallow water and expect the fish to hang around to have a go at your lure or bait! Generally, fish will move off to one side as a boat approaches in shallow water. After the boat passes, the fish may continue to swim away, stay where they are or return to their original location. Where you position your lines and how you manoeuvre your boat will determine how successful you are in getting your lure in front of these fish.

The real test of successful flatline trolling is to get your lure or bait into those tight restricted areas, like weedbed edges, along or really near to shore, close to islands and near standing or submerged timber—these are all difficult places to get a good trolling presentation. In heavily timbered areas or shorelines with submerged trees or rocks it is extremely difficult to get a lure close to shore without the risk of doing serious damage to your propeller or motor. Until the advent of planer boards the only way to deal with this situation effectively was to weave in and out from shore and to actively plan your approach to structure, points, weed bed edges, creek mouths and the like. Planer boards can allow you the freedom to explore these areas without the risk of damage to your boat or motor and without spooking the fish you are

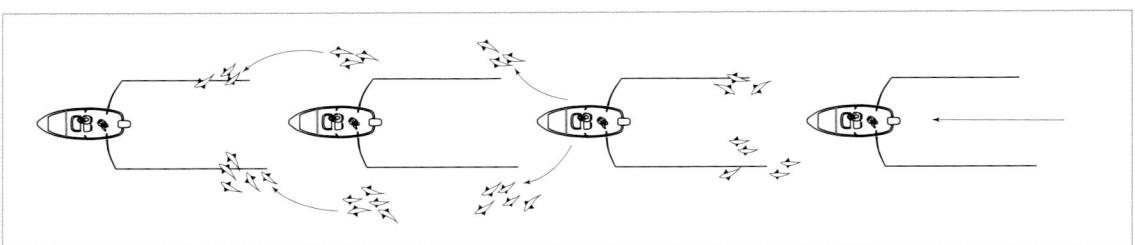

Figure 5.1: Trolling over fish in shallow water generally results in the fish moving off to one side or the other to get out of the path of the boat.

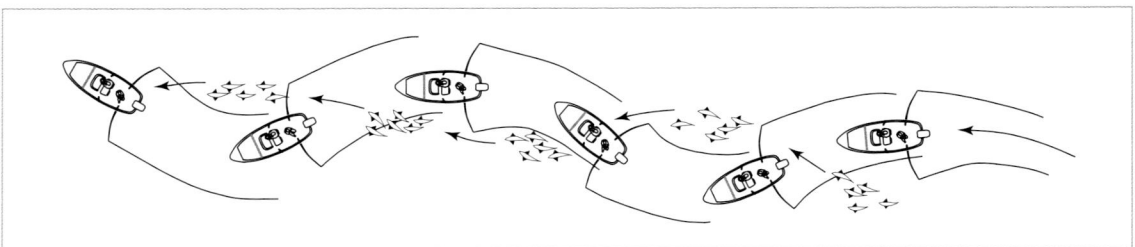

Figure 5.2: Working your boat in a series of S-curves, starting and stopping, and changing your boat speed are all valuable tools for helping get your trolled presentation over more fish!

Planer Board Trolling 47

targeting! Trolling will always be more productive if you incorporate more movement into your boat's path. Don't troll in a straight line, keep your boat working in S-curves and vary your speed.

IN-LINE PLANER BOARDS

These types of boards are generally small planers that attach directly to your fishing line; you don't need a towline to fish them. One of the more popular models is the Magnum/Offshore in-line board. These small side planers, like most boards, are sold in pairs; one runs to port (left) and the other to starboard (right). The board attaches to your fishing line ahead of the lure, and the lure is out away from the boat as far back as you wish. When a fish strikes, a release clip disengages and the small planer board flips loose, stops planing and slips down toward the lure.

In-line planers are similar to side planers but are usually a lot smaller. Generally made from plastic and foam they're usually small enough to fit in your tackle box. As the name implies, an in-line planer attaches directly to your fishing line. To use these boards, a lure or bait is set out behind the boat as you are moving forward. Once your desired setback is established the in-line board is clipped to your line. There are a couple of different methods for achieving this: one method uses a release clip (like a small downrigger release) and a snap. The line is run through a snap on the rear of the board and clipped into a release clip at the towing point or front of the board. The in-line planer is then set out at whatever distance off the side of the boat you choose. When a fish strikes, the line pulls out of the release clip and the board slides down the line attached by the snap. The board can be rigged to stop ahead of the hooked fish by using a swivel, bead and a leader. I'm not too keen on this method, as landing a hooked fish can be a real battle.

I prefer to run two release clips, rather than a release clip and snap. With this technique, you again set the lure out behind the boat and then clip both releases to your line. When a fish strikes, one release clip pops loose and the board remains attached by the second clip. Keeping tension on your line and the hooked fish, reel your line in and unsnap the board. You can then play the fish without the added weight and drag of the board. In-line boards can be run any distance that you choose, but generally 5–20 m is usually adequate. How far apart you want to spread your lures and how much boat traffic you have to contend with will determine how wide you run your boards.

On the plus side, small planers are very compact. Since you don't have to use a mast or stand pole, towline and releases, they are easy to use. On the negative side, since they are small and light they can be unstable in a small to medium chop. Also, because of mechanical limitations, you can only effectively send a small board out 10–15 m to the side of the boat and the amount of weight you can hang on these small boards is quite limited. Another problem is that fish sometimes pull these boards off to one side when they strike, and since there is

In-Line planer boards such as the Offshore board pictured are a great way to get your lures out into undisturbed water at a price that won't break the bank!

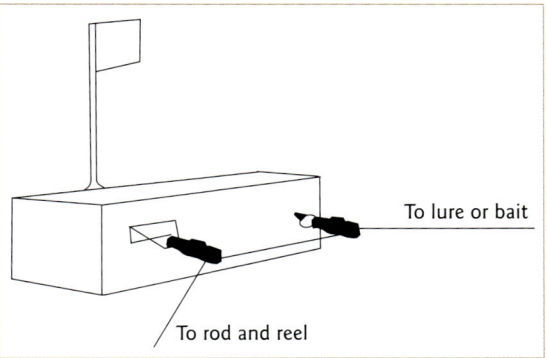

Figure 5.3: Rigging an in-line planer board: Once you have established your dropback (distance behind the boat) both release clips are fixed to your line. The amount of line let off your reel will determine how far the board will run out from the boat.

little resistance, you can miss getting hooked up. A further disadvantage of single runner boards is that when the towline goes slack they flop over on their side, which can cause tangles and false releases. Nor can you effectively use light line. For many anglers both of these facets can be a bit discouraging.

DOUBLE RUNNER SIDE PLANERS

Side planers are double boards usually constructed from wood or plastic. They can be fixed open or collapsible for easier transporting, and range in size from 0.3 m to 1.2 m long. This style of board is towed from a non-fishing towline (usually fluoro dacron from 100–200lb breaking strain) run off a mast or pole. The mast serves the purpose of keeping the towline up and stopping it dragging through the water. This style of board works much like a downrigger; instead of taking your presentation down it takes the lure or bait out away from the boat. Side planers also utilise a release clip, but unlike downrigger clips. Planer board clips have a hook that snaps onto the towline and looks something like a curtain hook. The release clip slides on the towline and its position on the towline is determined by how much line you let off your reel once the line is fixed into the release clip.

One of the big advantages with double runner boards is the ease of running multiple lines. I have run four lines off each side (port and starboard) without any hassles. The key to running multiple lines from one side is to stagger the length of your dropbacks. Put the longest dropback to the outside and any additional lines should have a shorter dropback. Be sure you have enough licensed anglers in the boat for the number of lines you want to run. Once a fish is hooked the line releases from the clip and you play the fish unencumbered.

Double-runner boards come in a myriad of sizes and they have the ability to ride very rough water without spinouts etc., they also do not flop over when there is slack in the towline. Their design is such that they ride quite perpendicular to the boat; even with a lot of line out they don't fall way behind the boat. When turning, they tend to keep travelling when single boards 'stall' and flop over on the inside of the turn. Also, because these boards tend to pull out at almost right angles to the boat, you can (if you have a sufficiently tall mast or stand pole) run 30 m and even more off to the sides without a lot of hassle. The biggest disadvantage with double-runner boards is their bulk, making them awkward to store. Some brands of side planers are collapsible, making them much easier to store. Although still bulky even when collapsed, I feel this is a small price to pay considering how well they work.

Lines trolled from boards react in much the same as flatlines when it comes to turns and speed. As you make a turn with the boat the outside lines will speed up and inside lines will drop or stall. This phenomenon can often trigger a strike from fish that have been following your lure. In heavy boat traffic you need to be very aware of where your lines are running. It can be very frustrating (not to mention embarrassing) to accidentally run your board into the bank, a tree or another boat!

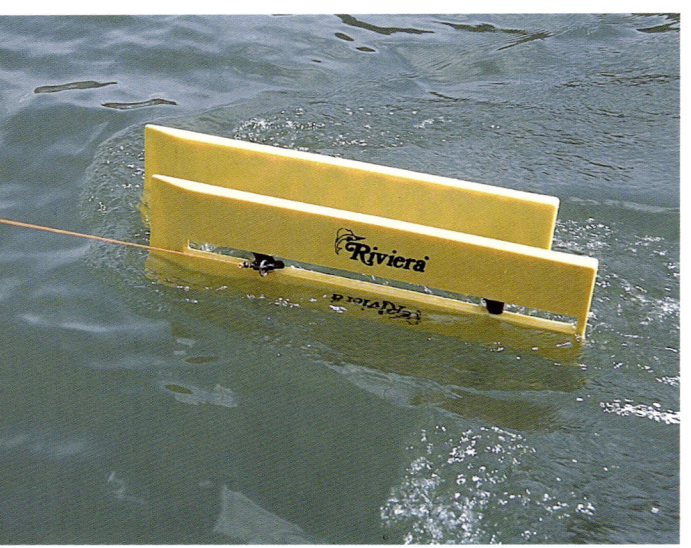

Double Runner Planer Boards by Offshore (Magnum) can handle very rough water conditions without getting out of control.

USING DOUBLE RUNNER BOARDS

Essentially there are three main elements to the side planer system:

1. A towline and/or reel to hold it;
2. A board and
3. A free sliding release clip, which holds your fishing line.

In addition, if you use a small boat for trolling you need a mast or pole to hold the towline up in the air so that it doesn't drag through the water. After that, all you need is your rod, reel and line, lure or bait and a rod holder of some sort.

Side planers need to have some means of

Masts for planer boards need to be high enough off the surface of the water (usually 1.5 m as a minimum) to stop the tow line from dragging through the water.

attaching the planer to the boat and a means of retrieving the board. Worldwide there are a number of manufacturers who produce side planers including Cannon, Big Jon, Riviera, and Magnum Fishing Products who distribute the Offshore range of in-line planers and release clips. Most current manufacturers produce masts or poles with either one or two reels for retrieving the towline. This mast should be mounted as far forward in your boat as practical. The further forward you can mount the mast, the better angle to the board, which helps keep the board abreast of, instead of behind the boat. One big advantage of side planers is that you don't have to reel in the board when a fish hits or your line releases. Simply set your drop back and set your line into a fresh release clip, snap it onto the tow line, let enough line off your reel to position the clip on the tow line and, you are ready to go! You need to have a good supply of release clips to take advantage of this technique. Release clips slide down the towline and stack up next to the board, but this doesn't cause any problems. The release clips can be retrieved when you next bring in the board. Keep in mind that because the boards are running in undisturbed water, and are themselves fairly unobtrusive, the amount of drop back from the release clip to your lure generally doesn't need to be as long as if your flatlining directly from your rod.

To use a planer board, start out (while the boat is moving) and set a board in the water (first on one side of the boat and then on the other side). Some

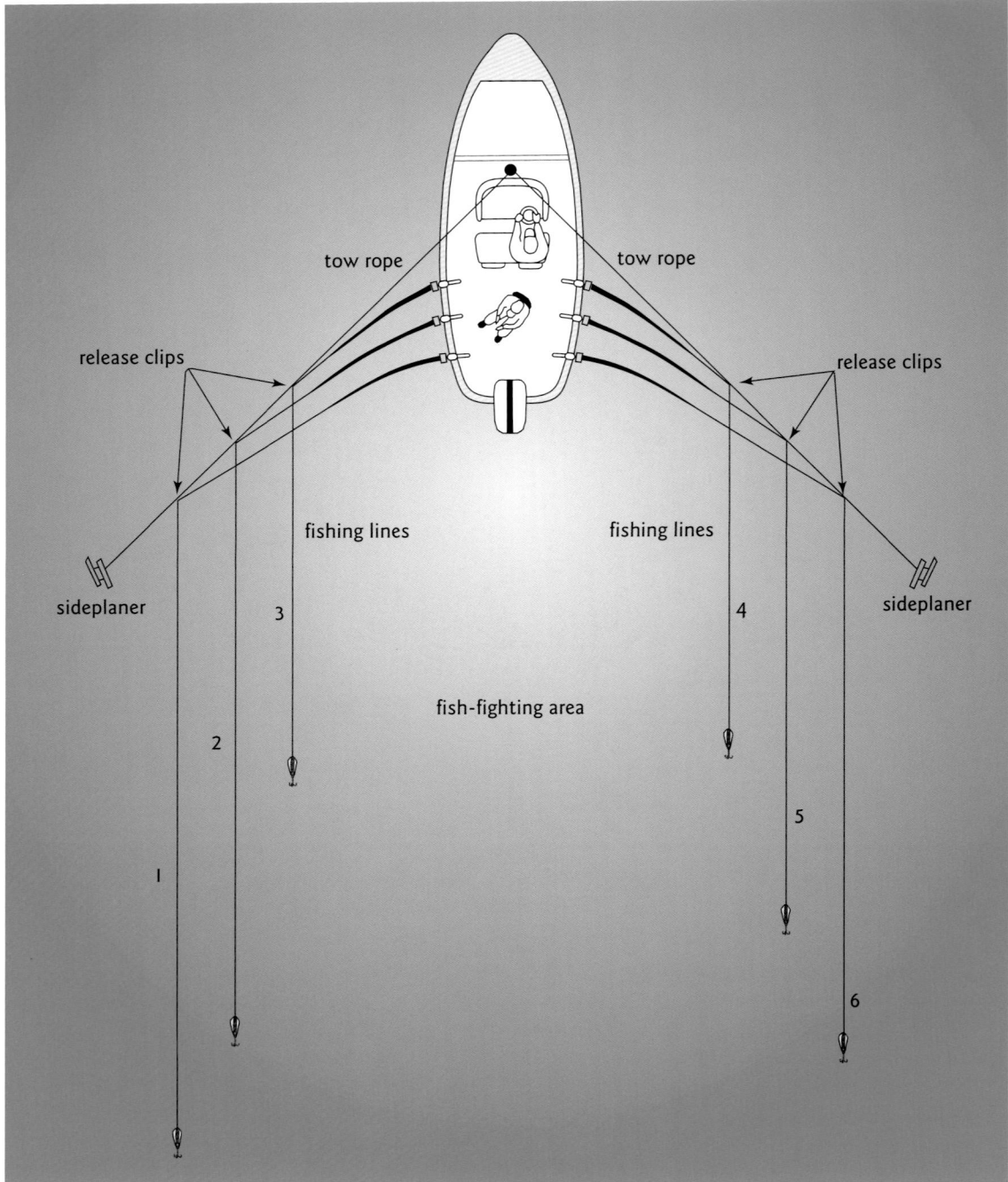

Figure 5.4; A typical set up of how planer boards allow an angler to run a spread of lures to target fish holding near shorelines or points.

brands of planer boards can be used on either port or starboard side exclusively and some are reversible. The mechanics of pull and the wedge-shaped nose design of the boards are such that as soon as the towline tightens up the boards immediately bite into the water and plane off to the sides of the boat. Once the boards are out the distance you want them you secure the towline and the boards simply ride off to the sides.

To deploy your presentation simply cast your

Planer Board Trolling 51

lure or bait to the side or drop it overboard and release line until the lure is as far out behind the board (dropback) as you want it, then snap the line into a release clip and attach the clip to the tow line. Now you're ready to send the fishing line off to the side.

Once again the mechanics of pull come into play. By releasing line off your reel the release clip, with your fishing line attached, automatically moves down the tow line and out toward the planer or trolling board. To stop the release clip from sliding all the way out to the board merely close the bail or stop the freespool in your fishing reel and the release clip stops where you want it. Set the rod in a rod holder. Your line is now out and working.

Towlines are usually quite long (sometimes as much as 30–50 m) so you are able to run more than one fishing line on a single towline. After you have rigged up and let out the first line simply repeat the procedure letting out a second or third line the same way. The only difference is you don't let the second line go as far down the towline as the first, or the third line as far down as the second. Now you are ready to rig lines for the other side of the boat. The same procedure applies. You can easily run three or four lines from a small 3–4 m tinnie. (Be sure you check you local regulations to determine the number of lines per angler that you can run).

If one line accidentally releases from the release clip (as it sometimes does if the tension is too loose) or if a fish hits and fails to hook up you can simply leave the planer board out and re-attach the line to a fresh clip. Just move the other lines over and down towards the board (you keep on trolling all the while) and reset the line again. In this way you can let a fresh line out without bringing the boards in.

When you hook a fish you can usually leave the boards and lines out and fight the fish in the open water behind the boat. Gradually you will accumulate a number of empty clips down by the board but this is not a problem as most boards are unaffected by clips stacking up next to them. Obviously you do need to have quite a few clips at the ready to avoid having to drag the board in just to retrieve clips.

In order for your lines to release without tangling they must be 'positioned' in a definite order. If that order is in any way changed so that the various lines overlap, or if the line overlaps with the tow line, it's wipe out time! Few boats are rigged (rod holders, mast or pole etc.) in the exact same manner. I can't give you a specific layout but I can explain the principles for positioning your rods in such a manner that when a fish is hooked it will swing to the back of the boat and will not (usually) cross over other lines.

To accomplish tangle-free fishing it is best to stagger your lines depth-wise (vertically) just as you do horizontally. For example, if you're running three lines on one board, the safest way is to

OPEN WATER/DEEP PRESENTATIONS
In this scenario fish can be targeted using a combination of techniques including surface lures, deep divers, lead core line, snap weights and downriggers. Be sure to run the outside line with the longest dropback and stagger any other lines to have a short dropback. The line to the lure depicts dropback not depth.

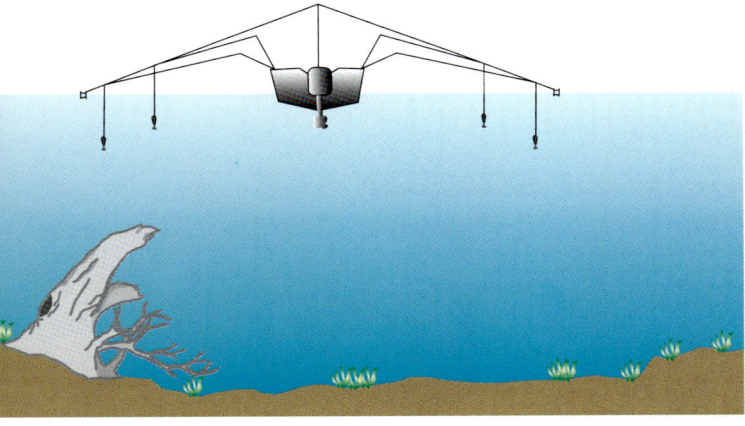

SPRING/SHALLOW WATER PRESENTATION
One big advantage with this technique is the ability to get your lure or bait into very shallow water without spooking the fish. In spring, early morning or late evening, fish cruising the shallows can be a prime target.

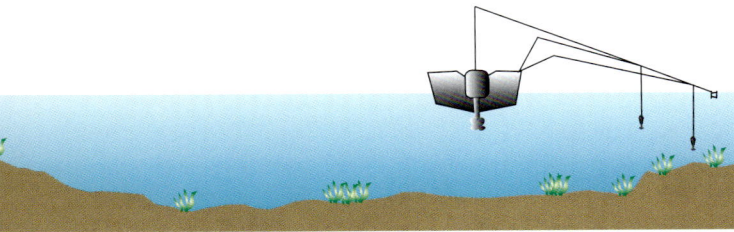

Figure 5.5: Planer Boards give you the option of targeting fish well away from your boat's path of travel, avoiding the problem of your boat spooking fish.

run the furthermost outline (the one nearest the board) the longest distance (dropback) behind the boat. Run your second line a little deeper, but not as far behind the boat. The third line is run deeper still and closer to the boat than the other two lines, etc. In this way when you make turns, or when a fish strikes, the lines swing in such a way as to remain spread apart and so do not cross over or come into contact with each other.

Remember, if the deeper running lures and lines were on the outside you would not be able to bring them back to the boat without fouling the line so you must ensure you set up carefully. Study the accompanying illustrations carefully; they could save you a bit of grief.

Tackle for Board Fishing

To produce consistent results while trolling with planer boards, it's imperative to know how much line you have out and to be able to reproduce those lengths. Counting passes of the level wind on overhead reels, marking the line in metre increments etc. are all ways of monitoring or determining how much line you have out. An easier, more consistent means is to use a line counter reel. Unfortunately most of the line counter reels I've owned or used are all a bit too big for the type of light line trolling that I most enjoy. After years of searching for a small line counter reel, I've finally found a viable alternative, and it is designed and made in Australia! The Tackle Tracka is not a line-counter reel but a device that clamps to your rod and gives you an accurate digital display of how much line you have out. In addition to the amount of line out, this device will also tell you the speed the line is moving in, or out, of the reel (great if you are casting and retrieving) and a whole host of other information, including an alarm for bait fishing. It will work with just about any rod and reel combination, thread line (spinning) or overhead (casting) with virtually no effect on casting performance. This is no toy, but a valuable piece of gear that can be used in a host of applications, as well as enabling you to produce consistent line lengths while trolling. If you happen to have a downrigger that does not have a line counter the Tracka can be a real bonus. Once you have set your dropback and have fixed your line into the release clip, the Tackle Tracka will display the amount of line you have out as you lower the downrigger weight.

Tackle Tracka. This compact little unit has an LCD display and will provide trolling anglers with a dependable means of monitoring the amount of line they have out at any time.

Your choice of rods and reels are mostly personal preference, but I can offer a few insights from my years of experience. Using double runner side planers you can run almost any kind of rod and reel. Unless I'm running an attractor, I often use a spinning reel to cast my lure and create a drop back before setting my line into the release clip. My personal preference is for a bait casting or overhead reel with a bait alarm or clicker. The clicker will enable you to free spool with clicker on, avoiding over runs and backlashes, as well as provide an audible indication of fish strikes.

Rod choice will again depend on which type of board you are using. With side planers, your choice of rod can be fairly wide, remember you have nothing on your line but your lure or bait. In-line boards are a different story however, keep in mind that you have a fair amount of drag created by the board attached to your line. I generally opt for a rod rated to 4–6 kg and 2.1–2.3 m long. An ideal rod will have plenty of backbone with a tip sensitive enough to provide an indication of a light strike or weed fouling the hook or hooks on your lure.

LINE CHOICES FOR BOARD FISHING

Line choice is another important consideration for board trolling. 2–3 kg line can be fine for use with a side planer, but with in line boards because of the drag created by the board, you need to run 4–5 kg line as a main line and use a lighter leader. Choose a line that is tough with minimal stretch and change your line often. Braided gelspun lines work well with this technique, but can create a bit of a problem with release clips. Because gelspun lines tend to be very slick they have a tendency to slide through or out of a clip very easily. Figure 5.12 below depicts a method of helping to combat this problem.

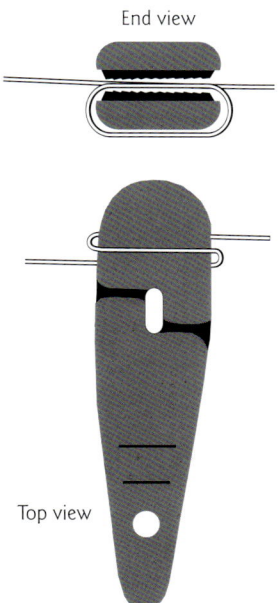

Figure 5.6: Using gelspun line in pinch pad style release clips.

Gelspun line or braids can be successfully used in this type of clip if the line is looped around the release a second time to prevent this slippery material from sliding through the pads. The diagram depicts how to rig these clips when using braided lines.

In Australia we are fortunate to have a wealth of lure styles and running depths to choose from. Both shallow runners and deep divers work very effectively when used on a planer board. Shallow runners that I've had good success with include Lofty's Lures, Wigston's Tasmanian Devils, Tillins, Tilsan, and Magnum/Lyman lures. In deep runners, the McGaths, Predatek, Halco, Legends and Australian Crafted Lures have all been productive.

Since speed is critical in any trolling application, the lures you run and how you mix or match them can have a real impact on how successful you'll be. There are lures or spoons designed for fast, medium or slow movement. There are even times when live bait like minnows, mudeyes or scrub-worms will out-fish any other offering. Experience and continual experimentation will help enormously with your success rate.

SOME USEFUL CONSIDERATIONS FOR PLANER BOARD FISHING

Listed below are a few points to note:

1. When using boards and release clips your tow point (mast or stand pole) needs to be a minimum of 1.5 metres off the water to avoid the towline dragging through the water.
2. Always make sure the line furthest from the boat has the longest dropback.
3. Keep deep diving lures on your inside (closest to the boat) line.
4. Always have a quantity of release clips at the ready.
5. Don't be put off by choppy conditions. Any wave action is imparted to your trailing lure; the rougher it gets, the more action on your lure.
6. Trolling along shorelines early morning and evening allows you to target big cruising fish with little water disturbance.

Once you have used boards for trolling you will soon realise that a number of factors come into play. Side planers should be used whenever water temperatures are at a level conducive to allow your target species to stay in, or at least temporarily go into, this temperature level. Each species of fish has a slightly different preferred temperature range. Knowing a fish's preferred temperature range will help in locating whichever species you're fishing for. A good temperature probe is an invaluable tool for board fishing.

Trolled lines run off planer boards can give you a great deal of choice when targeting specific species of fish. You can run shallow or deep diving lures, whatever you deem appropriate for the existing conditions. You can troll really hard pulling big lures for native fish such as cod or golden perch, simply by adjusting the tension on your release clips. You can even run leadcore line off a side planer by tying in a piece of mono to ensure the line releases from the clip. Snap weights are also an excellent choice for use with side planers as they can give you the option of getting down as well as out. The key to success with this technique is, like all methods of trolling, knowing how deep your lures are running at the length of line you have them set out.

Planer boards give the angler the ability to run multiple lures at varying depths and can also be used to troll lead core lines or snap weights to achieve greater depths.

Trolling with boards may seem fairly complicated, but in reality it's quite simple. To have the ability to cover a wide path of water with each trolling pass, get to fish in difficult areas, or get your presentation out away from the boat's line of travel can all be a big help in the troller's ultimate goal, getting fish in the boat! Planer or trolling boards are not limited to fishing for trout or salmon. There are obvious applications for practically all fish when they are up shallow and spooky!

CHAPTER 6

Getting Down
AIDS TO DEEP TROLLING

WEIGHTED LINES

LEAD CORE LINE TROLLING

Trolling with weighted lines like lead core is not everyone's idea of fun. Fishing with lead core means that most fish you hook will not give the same sort of sport that they would on light line. In most cases you are likely to find that if you have three to four colours of line out that a fish will have a real battle to drag that much weight around. The major advantages of lead core are that it is heavy and sinks, and has little or no stretch. The trolled depths you get down to with this line are directly related to speed, line diameter and the amount of line out.

Lead core line is produced by encasing a soft pliable lead wire inside a braided line, with a material like Dacron used for the outer covering. One common misconception that many anglers have about this material is that the greater the breaking strain the deeper the line will run. Not so! This line is like any trolled line in this respect. The smaller the diameter, the less resistance the line creates in the water, when you are trolling and the

Lead core lines are available in a range of breaking strains. Line in the 12–15lb Breaking strain is your best bet for trout trolling.

Reels and in-line planer board, all of which are suitable for lead core lines, snapweights and other deep trolling techniques

greater the depth you can achieve for a given speed. Lead core comes in a range of breaking strains of 12–45 pound. The only difference in the rating or breaking strain of this line is the different diameter of the Dacron cover; the lead wire is the same diameter for 12lb line as it is for 45lb line. The most common breaking strains of lead core line brought into Australia still remains 18lb and 27lb, though some importers are now finally bringing in 12lb and 15lb lead core. For freshwater trolling my preferred lead core line would be 12lb or 15lb breaking strain and 27lb or greater for saltwater use. Lead core line is colour-coded, every 10yd (9 m) the braided outside of this line has a different colour, making it simple to monitor how much line you have out. Lead core line is generally available in 100yd spools (10 colours) and 30yd spools (3 colours).

Tackle for Lead-Core Line Trolling

The choice of reels for using lead core effectively is limited to baitcasters or centrepin reels such as fly reels. Spinning reels just don't work with this material; the action of the line having to bend around the pick up point or bail will cause the lead core to break, resulting in a real mess. The size and type of reel you use should be determined by the breaking strain of lead core and the amount of line you want to load on the reel. Obviously, the smaller diameter lead core lines (12lb and 15lb) will allow you to load more colours (line) on to your chosen reel. I've used a lot of different reels over the years for lead line and have found that baitcasters like an Abu 5500 or 6500, Quantum Iron Baitcaster, Daiwa Millionaire. By its very make up, lead core is a bulky material. Whatever style reel you choose will have to suit your requirements to hold the maximum number of colours you may want to fish with. Fortunately, most of the major manufacturers of fishing reels produce a model of reel that is suitable for this technique.

When using lead core line for trolling, my preference is for a rod in the 1.95–2.1 m range (6ft 6in–7ft). Trolling rods with a medium tip and enough backbone to handle 4–8 kg line is ideally suited to this technique. Manufacturers such as Daiwa, Shimano, Loomis, Quantum, Strudwick and Shakespere all produce rods that are well suited to trolling with lead core lines.

Backing and Leaders for Leadcore Lines

Choosing a backing line for your lead core set up deserves a bit of thought and preparation. Good quality monofilament, braided gelspun line, and Dacron all make adequate backing material. My own preference for backing tends toward mono or braid as a backing material. Dacron is of course quite functional but tends to be somewhat bulky, and will use up more of your reel's available capacity. Braided lines with their smaller diameters will give you the ability to load more line onto a given reel. If you choose a braided line in the 10–15 kg (20–30lb) breaking strain for native species and 6 kg (12–15lb) for trout and salmon you can cover most scenarios.

Monofilament or fluorocarbon lines as leader material, in an appropriate breaking strain, are best used with lead core lines. The length of leader for lead core trolling should be determined by factors such as your target species, water clarity and the depth you want to troll. A general rule of thumb is to increase the leader length in clear water conditions when fish might be spooked by the line or boat. I have often had success running a leader as short as one metre, but in ultra-clear water conditions I've sometimes had to increase my leader to as much as 15 m (50ft) in order to consistently produce fish. Longer leaders have the advantage of getting your lure or bait further away from the boat, decreasing the likelihood of spooking your targeted fish. Fluorocarbon lines make a great leader material for almost any trolling application. I've used quite a few brands since they were introduced a few years ago and have since settled on the RIO range as reliable and tough with good knot strength.

Trolling attractors are sometimes the perfect marriage with lead core line. The action imparted by lead core line to lures and attractors is definitely different from how your lure or attractor works when rigged on mono or braid. In the right conditions the combination of lead core line with a dodger or bladed attractor is very effective.

Although some critics of lead core line trolling bemoan the disadvantages of this line I still believe it has its place in any troller's arsenal. It may well be a technique that you won't use too often, but it will have days when it can out-perform almost any other technique!

Wire Line

Little used today in Australia, wire lines are one of the early precursors to downriggers for getting a lure or bait deep while trolling. Wire lines are valuable for trolling because they are heavy and sink readily. Wire lines are generally made of single or multiple strands of Monel, (a corrosion resistant alloy) or stainless steel, both having their own advantages and downsides. Monel tends to be more expensive, but has the advantage of kinking less. Multi strand wire is a little easier to handle, but has its own problems when a burr develops. Single strand wire sinks more readily, so multi strand wire is less popular.

Wire lines sink at about 3 m (10ft) for every 30 m (100ft) of line out. Obviously if you want to get down to 9–10 m you will have to let out 90 m of wire. For this reason wire lines are often used in conjunction with weights (sometimes as much 400–500g) to eliminate having so much line out. In this technique line is let off the reel slowly by thumbing the spool as you release line. You can't freespool line out or you will end up with an almighty tangle! When you feel the sinker hit bottom, reel up a metre or so of line before putting the rod in a rod holder.

Due to the very nature of trolling with wire line it's necessary to use heavy tackle for this technique. You will need to run a heavy rod and large overhead reel with a star or lever drag without a level wind as the wire will cut into the level wind. This is more akin to some heavy saltwater techniques where you use your thumb or finger to help guide the line back onto the reel. It is also useful to mark the wire in some fashion to help determine how much line you have out.

I personally haven't used wire line for some years, as there is always an element of estimation or guesswork when using this technique. Modern trolling techniques, such as snap weights, can give you more reliable and predictable results with far less hassle and expense!

DIVING PLANERS

Diving planers have been around for a long time now, with many models pre-dating the introduction of downriggers in the late 1960s in North America.

Early planers were non-directional, and would only dive straight down. Early models included products like the Deep Six (originally made by Les Davis, then acquired by Luhr Jensen), Pink Lady, Dolphin and Jet Planer, all manufactured by Luhr Jensen.

Before the advent of downriggers, diving planers were an alternative to using wire line heavy weights or lead core lines to get a lure or bait down deep. To use diving planers is a fairly simple process. Simply attach the planer to your fishing line (1–1.2 m) in front of your lure and release line from your reel to determine how far you want the lure behind the boat. There are no weights to make the planer dive; the design of a planer is what determines how deep the planer will travel in the water. When a fish strikes your lure, a release on the planer trips allowing you to play the fish without the drag of the planer.

Jet Diver

The Jet diver is a fairly simple system that can accurately position a lure at depths of 10, 20, 30 or 40ft, with the added benefit of motion and colour. Each of the four models of Jet diver is depth specific, so to run at different depths requires another diver. Fitted with quick-change snaps, this injection moulded plastic diver only dives when motion is applied either by trolling or by current in a river. It does not contain lead weights to make it sink. Jet Divers float at rest, making them ideal for trolling in rivers for trout or native fish species.

To use a Jet Diver effectively, your depth sounder should give you an indication of the depth at which fish are located. Tie on the appropriate Jet Diver to reach your target depth, attach your lure 1–1.2 m behind the diver. Once you've completed this let out 30 m (100ft) of line and you're in business. Keep in mind that these diving planers are designed for fairly heavy line classes. The smallest diver the 'Jet 10' is designed for 12–15lb test line and the larger models 17–25lb test lines. As with all diving planers, diving capability and depth achieved are all affected by the length of line out, trolling speed and line diameter. Smaller diameter line such as braided gelspun lines will help achieve the maximum depth from any diving planer.

Fish Seeker

The Fish Seeker is an adjustable diving planer that works for a wide variety of fish species. This device can be used in both fresh and salt water and planes out when it reaches its set depth, which in turn means it can be trolled with lighter tackle. You don't need heavy gear (rods or reels) to use these planers. When a fish strikes your trolled lure the Fish Seeker will turn over and rise toward the surface so you don't have to fight the planer as well as the fish. If you happen to get a false release and a fish doesn't hook up, simply give the line some slack and the planer will dive back to its set depth.

Rigging the Fish Seeker

Using the chart in Figure 6.1 to determine what depth you want to troll at, simply attach your line to a snap swivel. Clip the snap swivel directly to a

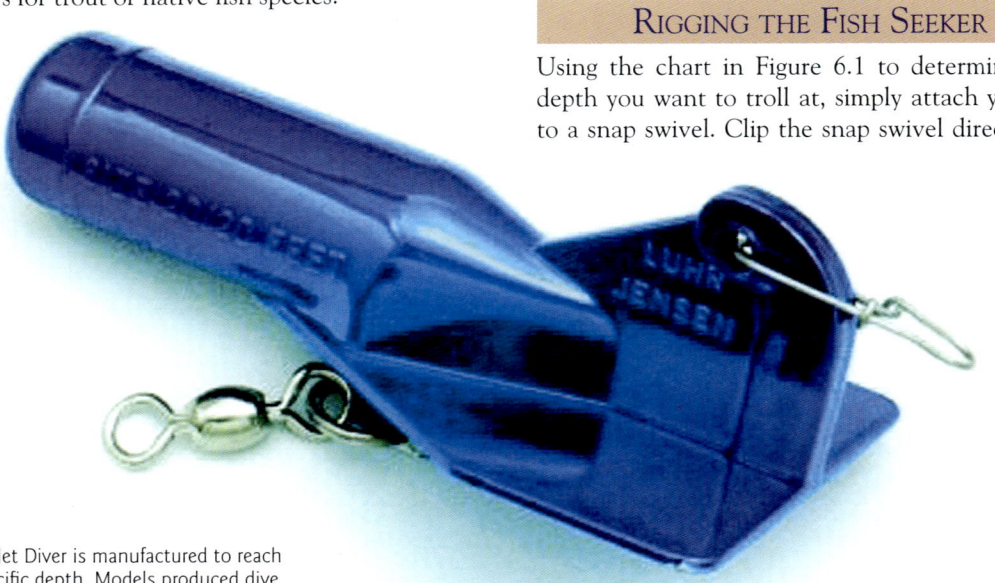

Each Jet Diver is manufactured to reach a specific depth. Models produced dive to 10, 20, 30 and 40 feet.

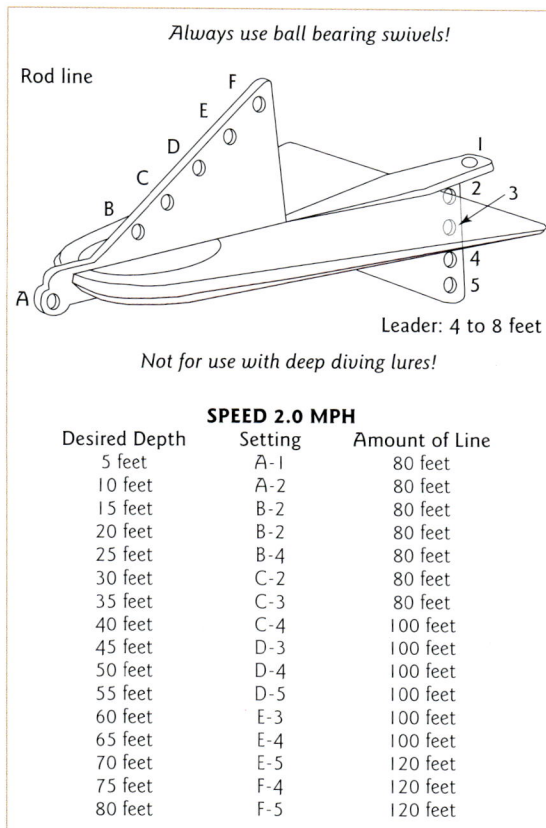

Figure 6.1: The Fish Seeker Depth Chart

SPEED 2.0 MPH

Desired Depth	Setting	Amount of Line
5 feet	A-1	80 feet
10 feet	A-2	80 feet
15 feet	B-2	80 feet
20 feet	B-2	80 feet
25 feet	B-4	80 feet
30 feet	C-2	80 feet
35 feet	C-3	80 feet
40 feet	C-4	100 feet
45 feet	D-3	100 feet
50 feet	D-4	100 feet
55 feet	D-5	100 feet
60 feet	E-3	100 feet
65 feet	E-4	100 feet
70 feet	E-5	120 feet
75 feet	F-4	120 feet
80 feet	F-5	120 feet

The Fish Seeker Diving Planer is a small adjustable planer designed to get your lure down to a range of depths.

lettered hole. Attach the snap swivel on your leader to one of the numbered holes. Tie your lure to the other end of the leader then release the amount of line shown on the chart to give the desired depth. Releasing more line in this instance will not cause the lure to dive deeper. Speed will however, affect the desired depth. Each increase of 1 mph will reduce the depth by 10 per cent.

DIPSY DIVERS

The Dipsy Diver is a directional diver, in other words in addition to diving straight down this device will also track to the left or right (of a straight path) and still get down deep. The big advantage for these devices is they allow you to cover a greater spread of the water column, allowing anglers to target prime fish holding areas. As an addition to downrigger lines they can add another technique that will help any angler to get more fish in the boat.

Like most diving planers, the directional Dipsy Diver is held in planing position by a lead weight. What's different with the Dipsy Diver is that the weight is moulded into an adjustable base plate. By rotating the plate to the right or left, it tilts the planing surface, causing the diver to move to either the port or starboard side when trolled.

After you've tied your line to the barrel swivel at the front end of the diver and have locked the adjustable trip mechanism in place, it is ready to fish. When trolled, water pressure against the diver's planing surface causes it to dive. Shifting the base plate weight to the left or right tilts the planing surface and results in the diver descending on a pre-determined angle. This multi-directional feature

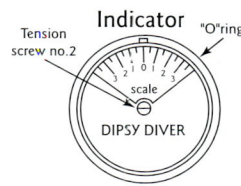

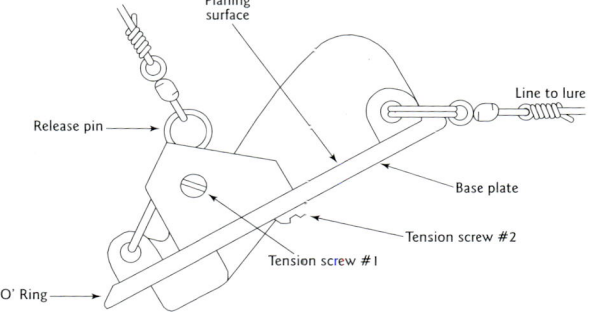

Figure 6.2: How the Dipsy Diver works

The Dipsy Diver comes in a range of sizes and colours.

Note: This chart uses 15lb Test Line

Line Out	25'	50'	75'	100'	125'	Side set
LES DAVIS DEEP SIX						
Size 0	13.3	22.1	28.9	32.8	35.9	N/A
Size 1	14.4	25.4	33.4	39.9	43.8	N/A
Size 2	15.6	28.4	38.0	46.0	53.9	N/A
LUHR JENSEN PINK LADY						
Size 0	11.3	19.2	26.9	32.1	36.0	N/A
Size 1	9.5	17.6	25.6	30.4	37.9	N/A
Size 2	14.6	27.3	36.0	43.1	53.8	N/A
LUHR JENSEN DIPSY DIVER						
Size 3/0	Not yet available					
Size 0	13.0	20.9	26.7	31.3	34.3	None
Size 0 Side set	7.1	13.6	18.7	22.8	27.5	3 (max)
Size 1	17.9	32.1	43.6	52.6	59.6	None
Size 1 Side set	13.5	24.2	32.2	39.1	45.4	3 (max)
LUHR JENSEN JET DIVER						
Size 10	6.8	11.8	15.4	17.7	20.3	N/A
Size 20	8.2	13.7	17.5	20.4	22.5	N/A
Size 30	8.7	15.0	19.8	24.0	27.8	N/A
Size 40	9.9	17.3	23.7	28.0	32.1	N/A

Figure 6.3: Depth Chart for Diving Planers

is possible because the base plate contains a moulded-in weight that can be shifted in degrees. Each degree of shift will cause a change in the angle of descent. A shift right or left as indicated on the base plate arrow (see diagram) will cause the diver to track straight, to port or to starboard.

Rotating the base plate past the edge of the numbered scale and then elevating your rod as high as possible can reach the maximum side movement of the Dipsy Diver.

Rigging a Dipsy Diver

Rig your Dipsy for fishing by allowing at least 4 to 6ft (1.2–1.5 m) of leader between the diver and lure or, when using a dodger, have at least 48in of leader between it and the diver. Adjust the No. 1 tension screw to hold the release pin in place while trolling, but be careful not to over-tighten it. Because this trip mechanism is adjustable, you'll be able to set it just right for the gear you're using. A slight tug on the line or strike by a fish should cause the pin to release.

The base plate is held in position by tension screw #2 that should be tightened just enough to hold the plate in position and still permit easy rotation for a new setting.

Note: the best results with a dodger are obtained from base plate settings 0 to 2.

Diving planers can provide any trolling angler with a number of positive benefits in addition to taking lures to greater depths. Diving planers provide attraction to any species of fish through their size, colour and action in the water. Lures behind diving planers are actually set a short distance behind the boat (both because of the total line out and the depth achieved). This means the trailing lure or bait is more easily affected by any boat movement (turns) or changes in speed.

Snap Weight Trolling

There is an alternative to using trolling sinkers and diving planers to gain extra depth on your trolled lure or bait! Snap weight trolling was developed in North America for targeting suspended or structure-hugging walleyes. To successfully target suspended or bottom-hugging fish, no matter what species, requires a special presentation of your lure or bait. Downriggers are without doubt the best tool for precise presentations at depth, but can be a costly exercise if you are trolling in very snag-infested water, and you hang up your bomb on a submerged tree or rock. A snap weight line can

Trolling in our deeper impoundments often requires the ability to get a lure or bait deep to fish suspended lower in the water column. Employing techniques such as lead core lines, diving planes and snap weights can all improve your chance of success.

easily be run in conjunction with a lead core line, flatline or downrigger. In addition, snap weight lines are perfect as planer board lines with either in-line boards or double runner trolling boards. This technique will prove a valuable asset for any dedicated troller, as it will allow you to troll more deep lines without a downrigger or lead core line.

At one time or another, we've all tried trolling sinkers of every size and description, but they all seem to have their drawbacks. If you run the weight close to the lure or bait you run the risk of spooking the fish, too far away and you can't land the fish because the weight is in the way; enough weight to get you deep and it takes the fun out of landing the fish.

Snap weights overcome these problems because they employ a pinch pad release (like a downrigger release clip) with a trolling weight attached. The real beauty of snap weights is that they allow the angler to choose how close or how far the weight is placed from the lure.

Basically the snap weight system works by allowing the angler to determine dropback, the distance the lure or bait is let out the back of the boat. Once this is established the snap weight is simply pinched open and attached to the line. The line, with weight and lure or bait attached, is then let out further behind the boat, simple and very effective!

A selection of snap weights can give the angler an inexpensive means of reaching different depths with trolled lures.

Snap weights can be almost any weight you choose. Commercially they are available in sizes from ½ to 3 ounces. Because snap weights are basically a release clip with a lead weight attached by a split ring, you can experiment with a range of weights. I generally carry ¼, ⅜, ½, ¾, 1 ounce and 1½ ounce weights for the type of fishing I do, but I have used heavier weights as well. Remember once you get the weight to your rod, you can unsnap the weight and play the fish unencumbered. If you choose to make up your own snap weights use a lead weight that is cylindrical or torpedo shaped to cut down on drag. The best known of the commercial snap weights is manufactured by Offshore Tackle Co. and imported into Australia by Magnum Fishing Products. Attach your choice of weight with a large split ring to the release clip, and you are ready for action.

Tackle Selection for Snap Weight Trolling

To troll with snap weights, most trout or native tackle will fill the bill admirably. To start out with this system, I'd suggest you look for a rod with a medium to fast action in the 1.8–2.2 m lengths. Ideally you want a rod that has a light enough tip to telegraph your lure's action, but still able to cope with dragging around a bit of a weight. Your choice of reel is really a matter of personal preference. Baitcasters, overheads, or spinning reels will all do the job, though my personal preference is to use an overhead reel with a line counter.

Line selection for snap weight trolling is fairly crucial. This technique does not require fine line diameters to help achieve depth; the weight takes care of this. Keep in mind that if you are fishing an impoundment with lots of snags your line is going to get a real beating. After you have landed a fish or been snagged, remember to cut a metre or two off the end of your line. Most of the stretch in mono or co-polymer lines occurs near the end of the line. Cutting a bit of line off and retying your lure or bait may save you the grief of losing a good fish. Look for a tough abrasion resistant line with low stretch and something in the order of 4–5 kg breaking strain. This line is your main connection to the fish; it doesn't pay to buy poor quality line. Over the years I've developed a real fondness for Platil Strong ST as a dependable trolling line. I've also used other brands like Platypus and Maxima with good results.

Snap weight trolling tackle needs to be heavy enough to accommodate the added drag created by the snap weight. Line counter reels such as the Daiwa model pictured are a great way of monitoring how much line you have out at any given time.

To make your presentation as accurate as possible you need some means of measuring the line you let off your spool. For years I played with counting revolutions of the spool on my reels, measured and marked my line in an attempt to get as accurate a reading as possible of my line out. I now use a line counter reel that makes the task of measuring line out a lot simpler. Be warned, they are fairly pricey. Another less expensive method of measuring line out is a clamp-on line counter that fits on your rod and measures line out after it leaves the reel. I've toyed with a couple of these little devices and so far, the one I like best is the Australian designed Tackle Tracka by Precise Angling. This device has an LCD screen and gives you a digital readout of your line out.

The main purpose of all these toys is to enable the angler to accurately repeat line out. If you catch a fish at a particular depth or drop back, you can repeat the process to get back to the strike zone. If you don't have a line counter, use a permanent marker to mark your line at regular intervals. I really can't overstress the importance of knowing how much line you have out with this technique. To avoid constant hang-ups, you need to know where your lure is in relation to the bottom or structure. Your sounder is an indispensable tool for depicting depth and structure. I use a Lowrance LCX-16 and I have found it to be an extremely accurate piece of equipment. The detail that most modern sonar units can give you is really quite amazing!

LURE SELECTION FOR SNAP WEIGHT TROLLING

We are very fortunate to have an enormous selection of lures to choose from. With the vast array of locally produced and imported lures, regardless of whether you are trolling for trout or natives, your choice is almost endless. Locally manufactured lures such as Lofty's Cobras, Tassie Devils, Tillins King

The Australian designed Tackle Tracka line counter has a range of functions including giving the angler a precise distance of line out.

64 Freshwater Trolling

Kobras, and the local minnow style lures such as Legend Minnows, Australian Crafted Lures, Minmins (Predatek) as well as imported offerings from Rapala, Luhr Jensen Quickfish, Rebel Minnows and Wordens Flatfish all work well with this technique. The snap weight system was developed to troll spinner baits and worm harnesses for walleye, but most minnow style lures work equally well. Just about any lure that will consistently troll on a flatline to 1 m to 1.5 m is usually a good choice. I would steer away from using this technique with deep-diving lures as trying to predict an accurate depth is just too complicated and unreliable.

Don't discount bait as an alternative with this technique. A big scrub worm on a worm harness like the Luhr Jensen wedding ring spinner or a walleye harness (yes, they do work on both trout and natives) can be deadly, especially in winter or spring when water levels are rising. Smaller attractors such as dodgers and small bladed attractors can be very productive when used in conjunction with bait, but with all but small dodgers it is very difficult to accurately predict depth.

When landing a fish you have hooked on a snap weight line, a steady retrieve is usually the best approach. Try to avoid the temptation of really setting the hook; you'll find that the fish generally do a good job of hooking themselves. A heavy jerk on the line may result in your pulling the hook out of the fish's mouth.

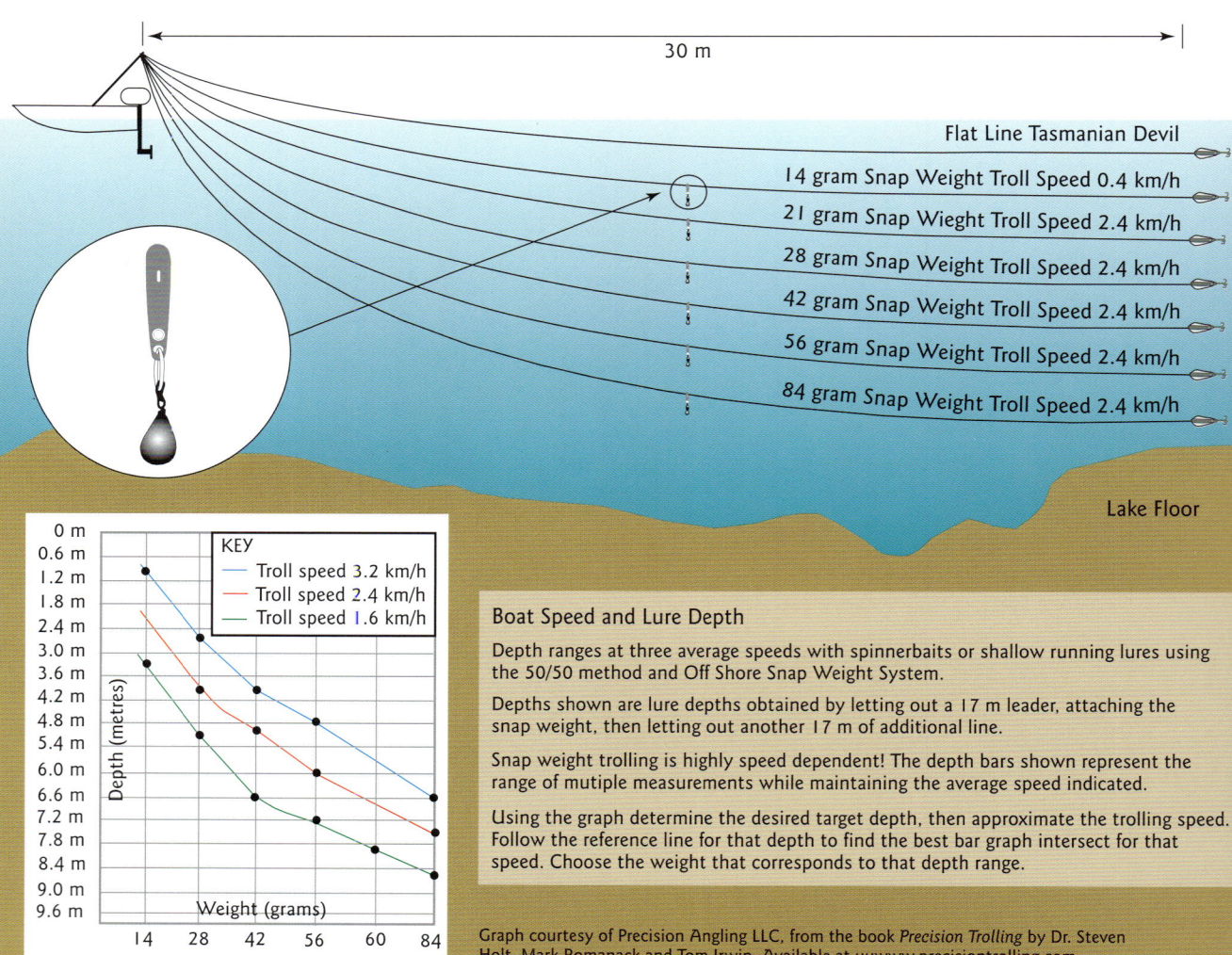

Figure 6.4: Snap Weight Depth Chart

Trolling during summer for trout and salmon can be frustrating if you don't have the ability to get your lures deep. As water temperatures rise fish will seek depths that can provide the right mix of temperature, oxygen and food sources.

BOAT SPEED AND LURE DEPTH

The most widely used application for snap weight trolling is when you choose to take a normally shallow running lure to a greater depth. If you want to get the most out of this technique, remember that snap weight trolling is highly speed dependant. The slower you go, the deeper the lure will run. This also applies to the amount of weight you use; heavier weights are also going to take your lure deeper. Boat speed in the vicinity of 1.5 km/h is a good starting point. The standard trolling method with snap weights is called the 50/50 system, developed by professional walleye anglers to help standardise leads or dropbacks and to more accurately predict depth. It relies on your main trolling line being let out behind the boat 50ft (approximately 15 m), the weight snapped on and another 50ft (15 m) let out. The graphical chart is a good a starting point, and is intended as a guide to get you started. With the number of variables in this system you need to consider carefully what presentation you want. The information in the graph is accurate to within about 1–1.5 metres. Your own experience will soon have you working out depths fairly close. Remember that your presentation will run slightly deeper than the snap weight.

ADVANCED APPLICATIONS

My personal favourite use for snap weights is running them off a planer board. By using snap weights on my planer board lines I can increase my coverage of fish by having a line that is not only

Australian native species including golden perch and Australian bass can often suspend in depths from 2–10 metres. Getting your presentation down to these depths is vital to ensuring consistent results.

out in undisturbed water, but down as well. This technique is often very productive if your target fish are in clear water and are spooky. Native fish such as golden perch are prime candidates for this technique. To target fish that are suspended it is very handy to be able to run a couple of snap weight lines in addition to your downriggers.

Another application for snap weights is to use them as bottom bouncers by placing them closer to the lure. This technique is a common approach for targeting bottom-hugging fish, and involves bouncing your snap weight off the bottom to attract fish to the trailing lure or bait. This is not a technique for the faint hearted, you can sacrifice a bit of gear to snags etc. but you can also catch big fish! Again this is another technique developed in North America, but it also has enormous potential for our fishing, for natives as well as trout!

CHAPTER 7
Trolling Attractors

TROLLING ATTRACTORS FOR FRESH WATER

Attractors for trolling come in a myriad array of shapes and sizes, not to mention just about every colour you can think of. Almost anything can act as an attractor when fished from a downrigger. Things such as downrigger weights and even sinkers can act as attractors. The range of attractors available from tackle shops can be quite intimidating if you are not familiar with them. Basically attractors can be divided into three main styles or types: rotating blades (cowbells, ford fenders etc.), dodgers and flashers. I'll give you an explanation of the different types and a few hints on how to best utilise them.

DODGERS

One of the most versatile and effective trolling attractors available to anglers today (and a personal favourite) has to be the dodger. I can recall seeing homemade versions of these attractors in the late 1960s on Lake Michigan in the United States and thinking to myself that these things would probably frighten off every fish within miles. How wrong I was! They had exactly the opposite effect and are now standard fare for most charter boat skippers working the Great Lakes in the US. Keep in mind that conditions and rigging for the Great Lakes are considerably different from conditions here in Australia. I've always felt that the translation of the technique to Australian conditions lost a lot in the adaptation. I'll see if I can clarify a few misconceptions and myths.

Firstly, most attractors developed early on for the American market were designed for chasing big salmon and steelhead. Standard tackle in those days was a medium heavy rod and big star drag overhead reels spooled with 10–15 kg (20–30lb.) monofilament. Awesome tackle for chasing freshwater

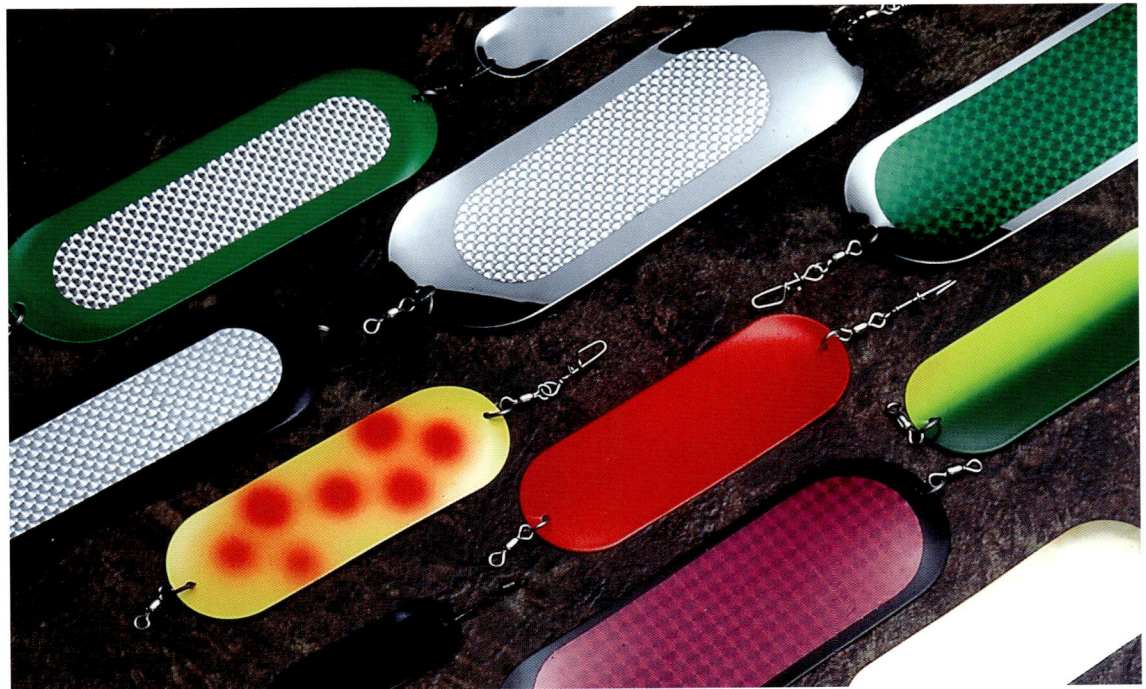
Dodgers are one of the most effective attractors available to anglers today. They can be used in conjunction with lures, flies or bait.

Trolling attractors such as the large dodgers depicted, make a very effective tool for running off the downrigger bomb as an attractor.

species, and attractors like dodgers were scaled accordingly. Huge dodgers up to 400 mm long (16") were not uncommon for deep trolling although smaller dodgers in the 6" (150 mm) to 10" (250 mm) were more the norm. Back then downrigging was still in its infancy and the few people using downriggers relied mainly on homemade units to do the job. Lead weights and diving planers were used extensively to get lures and bait down to the fish. Water conditions and clarity were pretty ordinary as well as the pollution of these lakes was probably at or near peak in the late 60's to early 70's. Fortunately for anglers in the US this problem with clarity and pollution has been addressed and conditions have improved dramatically. Many successful anglers are now applying techniques more akin to the way we fish in Australia, lighter tackle and a more sporting approach!

Dodgers are basically a single thin blade (0.4 to 1.0 mm thick) stamped or formed from sheet brass or copper. As part of the manufacturing process they generally have a lip of rolled edge on either end to create their unique action. Sizes are determined by the overall length of the dodger and can vary from about 50 mm long up to monsters 400 mm long. An appropriate size for our freshwater conditions would be in the 50–150 mm long range. Dodgers usually come rigged from the maker with either solid rings or split rings and swivel to help eliminate any line twist. In my experience dodgers create less line twist than almost any other style of attractors. They also create much less drag or pull than other styles of attractors.

Though I've heard countless anglers bemoan the fact that a 200–250 mm long dodger is too big for our style of fishing, I would suggest that a couple of these big monsters do have a place in our tackle boxes. Used as attractors off the downrigger bomb they can be extremely effective.

Dodgers impart an erratic darting action to

any lure, bait or fly but also create attraction by vibration and flash. The correct trolling speed for dodgers is achieved when the attractor develops a side-to-side swaying motion. You can check this by placing a rigged dodger and lure in the water while moving as slow as your boat will go. Slowly increase your speed until the dodger has developed the right action. The amount of action imparted to a trailing lure or bait by a dodger is determined by the size and weight of both your dodger and lure or bait. The speed at which you choose to troll will also impart the amount of action you get on this combination.

There is a vast assortment of sizes and finishes available for dodgers bought from tackle shops and mail order businesses. Dodgers are available with painted, powder coated, plated (gold and silver), prism tape and metallic or metal-flake finishes. Brands that I've used and had good results from include Luhr Jensen, Sep's Pro fishing and Magnum. Over the years I've probably experimented with just about every size, brand and colour of dodger available as well as making and painting my own. Like most anglers I guess I have my own favourite colours, amongst them would be plain brass or gold, fluoro pink, chartreuse, black and gold or black and silver. These colours are the ones I would probably choose first when prospecting new waters. We are fortunate in Australia to have a manufacturer that produces a range of products designed and made here for our conditions. Magnum Fishing Products make a very good product with a great range of colours and at a price that won't break the bank.

Rigging dodgers and lure combinations are critical to their success, but by keeping a few simple principles in mind you can be assured of consistent results. Try to match the size of dodger to the fish you are targeting and the line class you are using. It doesn't make sense to try and use a 300 mm dodger on 2 kg line chasing fish of less than a kilogram! Remember that the further you have your lure away from the dodger the less action it will impart to the lure. If you are running flies or bait behind a dodger the length of leader should be around the 200–300 mm mark behind the dodger. Floating minnow style lures should be rigged with a leader of 400–600 mm in length. Tassie cobra style lures (Tillans, Johnson, Lofty's or Wigston's) can run leaders between 400–1200 mm as they have a very strong action on their own. These lengths are generally a good starting point, but I would strongly encourage you to experiment with different lengths as weather and water conditions or clarity change. The drawings in this article include rigs for downrigging in different applications.

Dodgers come in a vast array of sizes and colours. In addition to the locally produced models from Magnum, several companies in Canada and the US produce dodgers that are suitable for our conditions in Australia.

Flashers like the models shown are great attractors for trout and salmon. The smaller models are fine to troll on 2–4 kg lines. Like dodgers the larger sized flashers make a great attractor off the downrigger bomb.

Flashers

Flashers are very similar to dodgers in general appearance with only slight differences. Flashers are turned up on one end and down on the other end, much like a dodger. Unlike dodgers, flashers have the appearance of an elongated flattened S shape, sometimes with fluting or ribs stamped into them. Some flashers are manufactured with a taper (wide at the front, tapering to smaller at the back) to enhance their action. Some large flashers come factory rigged with ball bearing swivels on the head or front to eliminate any line twist. These attractors come in a range of fairly large sizes from about 150 x 25 mm to monsters 400 mm long, and are made from either plastic or metal. Colours are similar to the broad range available in dodgers. Some of my favourites include the Alaskan Eagle and Abe and Al from Luhr Jensen and the range of Bechhold & Son Flashers from Magnum Fishing Products.

Flashers attract fish in much the same way that dodgers do by creating an erratic action plus flash and vibration to trailing lures or bait. Flashers have the same side-to-side sway as dodgers with one exception—flashers are actually designed to make a full 360 degree rotation. When you have the right speed for a flasher it will completely turn over (rotate) every four or five sways, even without a change of boat speed.

The biggest drawback to using flashers is that all but the smallest sizes are just too big for trolling with 2 to 4 kg line. Just keep in mind that these things were designed for 30 kg King salmon and big steelhead and you can see why the size is all wrong. I have used a small (150 mm x 25 mm) flasher on 3 kg line with good results, but any larger is just too big. Larger sizes, such as a 200–250 mm flasher, are my first choice as an attractor off the bomb when downrigging. If you use a flasher or dodger as an attractor off the bomb be sure you run a heavy ball or bell sinker on a short leader (300–400 mm) from the end of the attractor (see drawings) to stop the flasher from spinning out of control. Remember to stack your main line and lure at least the length of your flasher and leader above the bomb.

Bladed Attractors

Of the many styles or types of trolling attractors available to Australian anglers the most widely

Bladed attractors can provide a lot of flash and vibration to attract fish. How much drag they create on your trolled line will depend on the number of blades as well as their size and shape.

Ford fenders, cowbells and other bladed attractors are an effective tool for the troller, but can create a lot of drag on your line. This style of attractor also works well when used as an attractor run off the downrigger bomb.

known would have to be cowbells and ford fenders. Both of these attractors are part of a family of attractors commonly referred to as bladed attractors. These attractors all have a few things in common: they generally have individual blades that revolve around either a rigid wire or flexible stainless steel cable and they almost always have either a plastic or metal keel to help eliminate line twist. The array of configurations of blades is staggering with some American models having as many as seven or eight blades. With the exception of ford fenders, which have their own unique shape, bladed attractors can utilize almost any style of blade. Some of the most common models include Colorado, Indiana and Willow Leaf blades. Some of these attractors feature different size blades of the same style and some even combine different styles and sizes on the one attractor. As with most other types of attractors the range of colours and finishes is almost limitless.

Blades for this style of attractor are usually stamped from brass or copper. I've seen some huge aluminium blades made by Luhr Jensen, called Jumbo Aluminium Light Trolls, with blades approximately 200 mm long. The idea behind the aluminium blades is that they supposedly are much easier to tow with less resistance, with their huge size they certainly should provide a lot of attraction.

The best-known bladed attractor in Australia would have to be the ford fender. Used in conjunction with live mudeyes they can be one of the best fish producers for Australian conditions. Given this fact I still rarely ever use ford fenders, mostly because they create so much drag on the line. To use these things effectively calls for heavier tackle than I prefer to use for trolling. On the rare occasion that I use ford fenders or cowbells (these two names are actually registered designs of Luhr Jensen, though they've become synonymous with this style of attractor) I tend to run them as attractors off the bomb when downrigging.

One of the newest innovations in bladed attractors is a product called T.H.E. Flashlites. These lightweight attractors have Mylar (a type of flexible plastic) blades covered with prism tape that rotate on a flexible wire. Unlike most bladed attractors, some Flashlites don't have a metal or plastic keel; instead they utilize a small split shot crimped on to a tag from the main wire. T.H.E. Flashlites weigh almost nothing, but you still know they are on the end of your line. I've found them to be excellent for use as an attractor for all types trolling.

Of the many styles of blades used for bladed attractors my personal favourite is the willow leaf blade. I've used these things for a lot of years with great success and they are available in a huge range of sizes and colours. Several manufacturers make these attractors including Wonder, Luhr Jensen and Gibbs. The brand I've used most recently are distributed by Pieces Tackle in Melbourne and are available in a range of sizes down to a number 3 (about 25 mm long) which is excellent for light line trolling. When using very small attractors (both dodgers and bladed attractors) I frequently run a small down under sinker or lead keel in front of them to get a little extra depth. You'll know when you've got a good trolling speed (very slow) by the slow regular pulsing action of your rod tip.

Big, heavy bladed attractors are going to create even more drag on your line so you may want to consider an alternative. Your selection of attractor should, at least in large part, be determined by water and weather conditions. On bright sunny days with little or no wind I would think twice about running a big flashy attractor as it could spook the fish. In low light (heavily over-cast days) with a good chop on the water that same attractor could just be the ticket!

Leader length to attractor is often a matter of personal preference. As a general rule of thumb the shallower and clearer the body of water you are fishing the longer the leader you should use. In very dirty water after heavy rain I've used leaders as short as 1.2 m, yet in New Zealand's Lake Taupo I've had to run leaders 10 m long to consistently produce fish.

Length of leader from attractor to the lure will again depend on what you choose to run. A good starting point for lures is around the 600 mm (0.6 m) length, but being prepared to experiment according to conditions is what will improve your catch rate.

DOWNRIGGING WITH ATTRACTORS

Choosing an attractor for downrigger fishing is really not all that difficult if you keep a few basics in mind. Obviously the deeper your lure or bait is in the water the less light penetrates the water. In my experience and from the research I've seen the colours that appear to be the most visible to fish in deep water would have to be gold, silver and chartreuse (fluoro yellow). In deeper water (10 m or more) one of these colours is usually a good starting point.

Bladed attractors, dodgers and flashers are all good attractors for downrigging. The harder pulling attractors like ford fenders and cowbells will force you to run fairly tight release clips to counteract

Trolling attractors can be fished with a range of techniques including downriggers, lead core, wire lines and snap weights. Though normally associated with deep trolling techniques they can also be very effective on flatlines.

their drag. Dodgers and flashers create much less drag and as I've mentioned earlier are my first choice for downrigging. Big flashers, cowbells and ford fenders are excellent as attractors when run from the downrigger bomb. Countless times I've seen fish on the sounder move up as much as 10 m to have a look at these things. Needless to say if your lure is in the same vicinity you can anticipate some real action.

TROLLING ATTRACTORS WITH BAIT

One of the most effective yet neglected techniques for trolling with attractors is the use of bait! On lakes or impoundments where it is legal to use bait this approach can be one of the most productive methods available to the freshwater troller, especially in spring! Big brown trout that spend a considerable amount of their time cruising and scavenging along the shores of impoundments in spring are a prime target for this method. Fish such as gudgeon, galaxids, smelt, and small redfin or any other predominant forage fish can be a real turn on for trout. Dead or injured baitfish are easy targets for a hungry brown in shallow water. For the diehard lure dragger such as me this may seem to pose a bit of a dilemma, but I can assure you the results are worth the effort

Baits such as large minnows or whitebait can be deadly behind a dodger or bladed attractor. If you run a live minnow, rig the attractor 400–600 mm in front of the bait with the minnow hooked through the lip to help stop the bait from spinning out of control. For dead baits, rig the dodger 200–300 mm in front of the bait, run a single hook through the bait's nose as the tow point (pictured in the drawing) then rig a short leader (approx 100 mm) with a treble hook snelled to the single hook. Hook the treble in the side of the bait, being careful to leave a little slack in the line between the single and the treble hook: this will enable you to troll the bait by towing it from the nose. If the line between the single and treble is tight it will cause the bait to bend, resulting in a lot of spinning or rolling. If you want to troll a large minnow (dead bait) without an attractor, rigging in this manner can be very effective.

Another springtime bait trolling option is the use of scrub worms behind an attractor. Most of the varieties of bladed attractors will work very well for this purpose as well as small to medium dodgers. Rig the scrub worm from 200–400 mm behind the attractor. Like trolling mudeyes, a slower presentation often works best. I have also found that some of the plastic worms used by Canadian and American steelhead anglers for drift rigging

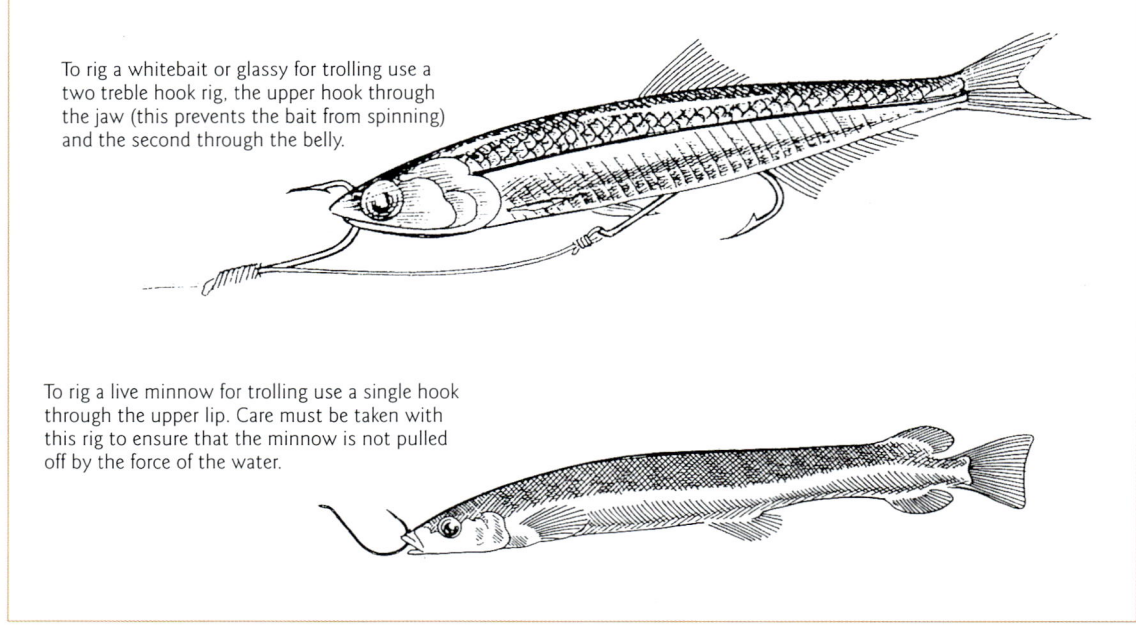

Figure 7.1: Rigging a whitebait or minnow for trolling.

Trolling Attractors 75

Bait trolling rigs such as the Bechhold & Sons "Fish Catcher" (distributed in Australia by Magnum Fishing Products) make rigging a minnow for trolling very easy.

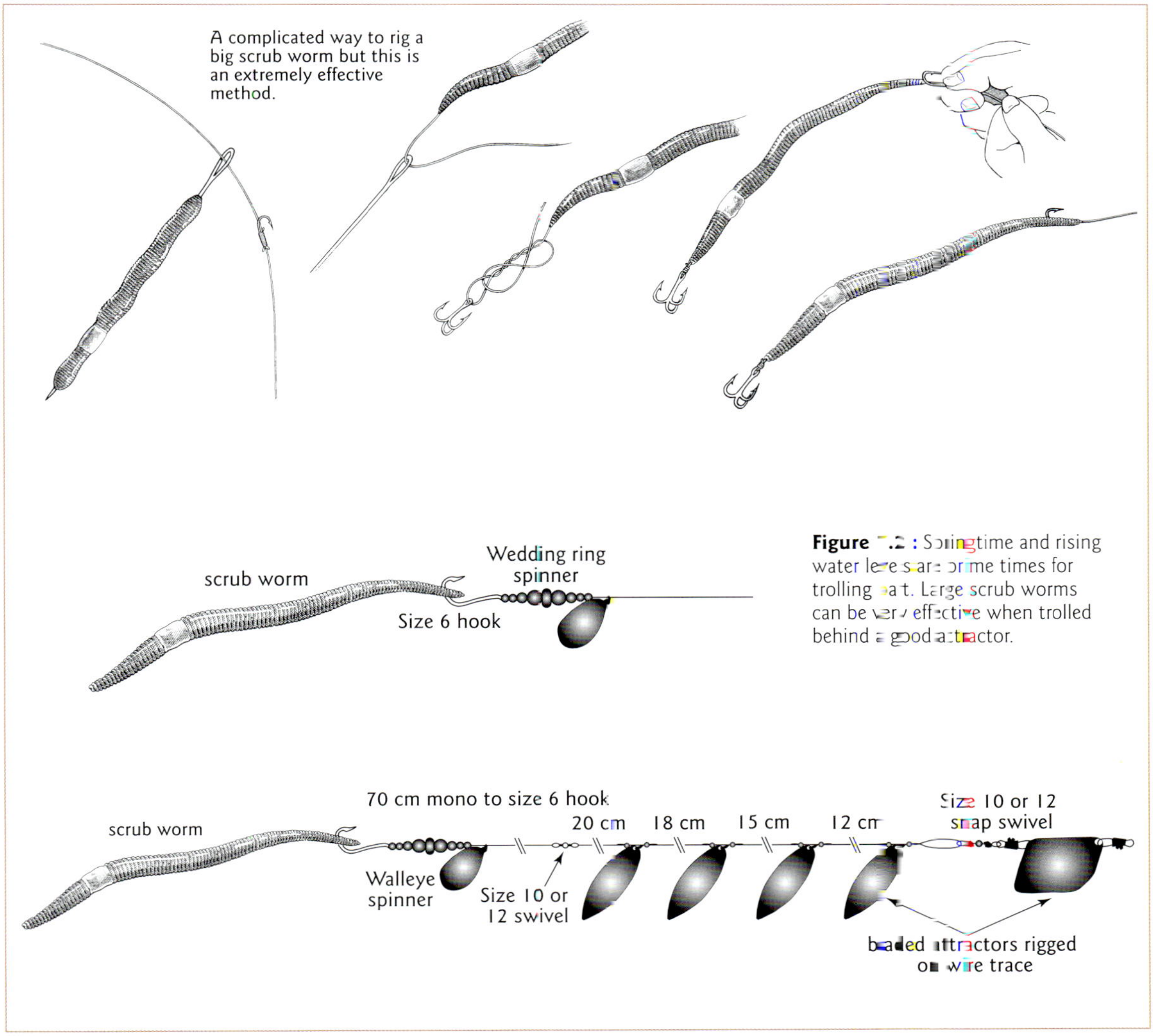

A complicated way to rig a big scrub worm but this is an extremely effective method.

Figure 7.2: Springtime and rising water levels are prime times for trolling bait. Large scrub worms can be very effective when trolled behind a good attractor.

scrub worm — Wedding ring spinner — Size 6 hook

scrub worm — 70 cm mono to size 6 hook — 20 cm — 18 cm — 15 cm — 12 cm — Size 10 or 12 snap swivel

Walleye spinner — Size 10 or 12 swivel — beaded attractors rigged on wire trace

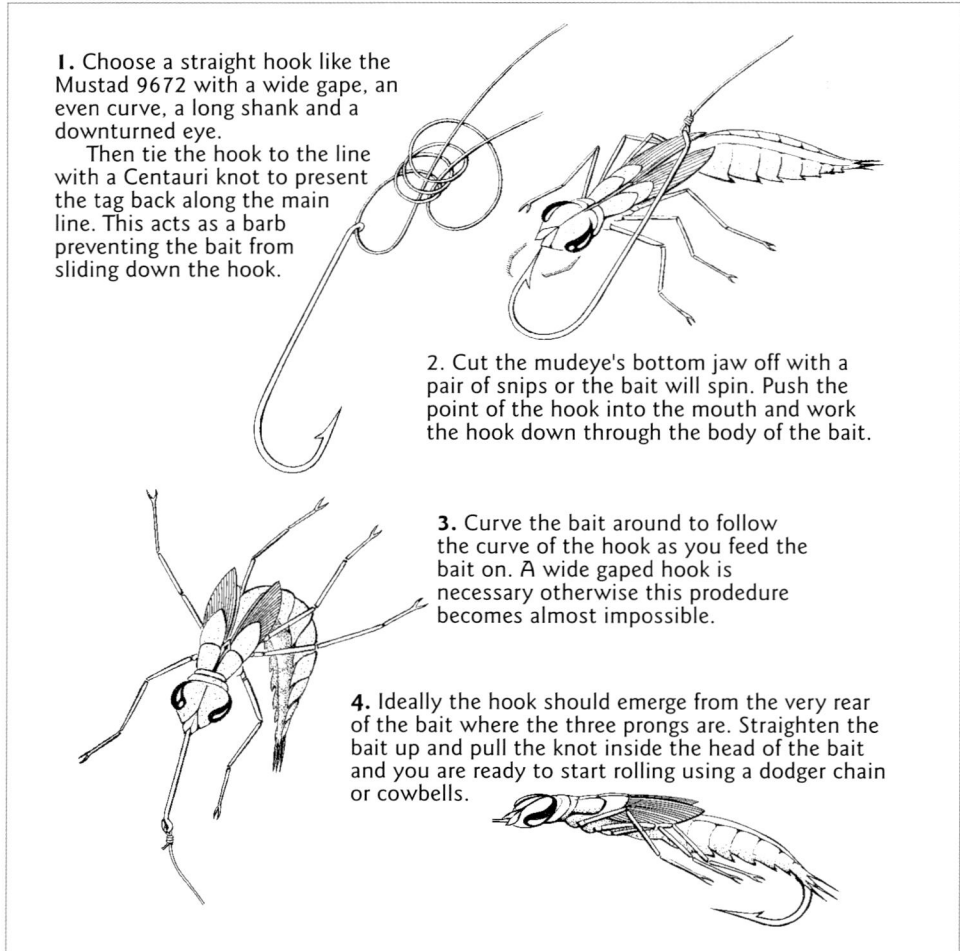

Figure 7.3: Trolled mudeye rig.

also make a great presentation behind an attractor for downrigging. My best success has come from a bright fluoro pink plastic worm behind a small dodger. The plastic worms range in size from 100–200 mm in length.

Trolling mudeyes (dragonfly nymph) has been one of the most successful bait rigs used in Australia for freshwater trolling. Though somewhat difficult for the novice to rig, perseverance with rigging this technique will pay off handsomely. The mudeye is best trolled behind an attractor such as a ford fender or cowbells, which transfer their pulsing action to the mudeye, making it irresistible to most trout.

The use of attractors should be a part of a successful troller's tactics. They are not always the answer when trolling, but they have been proven to be so effective they are bound to improve your catch!

CHAPTER 8
Downrigging Techniques

DOWNRIGGER DESIGN AND CONSTRUCTION

Downrigger design and construction does not have to be complicated, as the main function of this equipment is to raise and lower the downrigger weight.

A manual downrigger consists of a spool to hold the wire, a boom or arm, body, pulley, and crank or handle to wind the wire cable. Whether the crank or handle is on the side or top of the reel doesn't really make much difference to the function of the downrigger, and is more closely related to what you get used to.

The choice of what downrigger you ultimately decide to purchase is partly personal preference, but should be governed by more than price. A good manual downrigger doesn't cost a fortune as boating and fishing equipment goes. Some top of the line freshwater casting and spinning reels can cost twice as much as a good downrigger. Not a huge investment when you consider the years of use you should get from this equipment.

The single most important thing to consider when looking at the purchase of a downrigger is the construction. Keep in mind that this equipment needs to be very robust. If you consider the amount of leverage created by the weight being suspended from the end of the boom or arm, you can start to appreciate the forces acting on the downrigger. If you plan to do some serious deep (50 m) saltwater trolling you might need to use a weight of 9 kilograms (20 pounds). Obviously the downrigger you choose needs to be strong enough to cope with this kind of use. Consider carefully the type of materials the downrigger is constructed

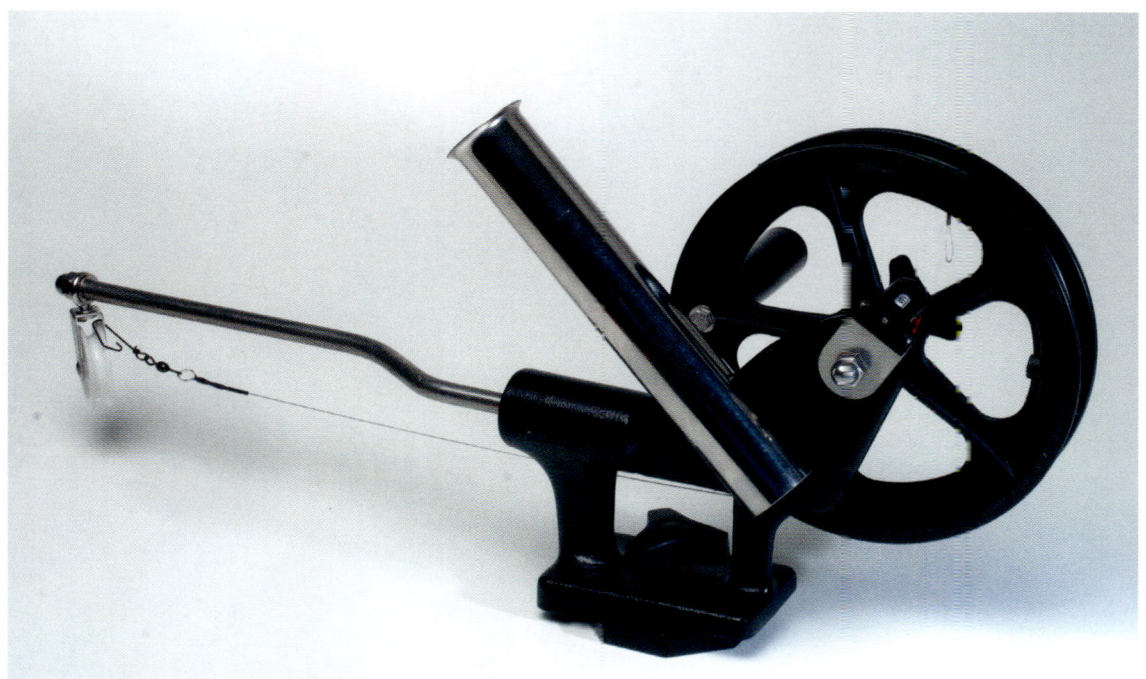

Magnum "Maxi" Downrigger. This Australian made product has been around for a lot of years now and has proven to be both durable and affordable.

The Fathom Master downrigger by Penn Reels. Like most of the Penn products this downrigger is robust enough to stand the rigours of both salt and fresh water.

from. Materials like aluminium, moulded nylon, polycarbonate (Lexan etc) and stainless steel are appropriate. Look carefully at how the unit is put together—corrosion of the parts could cost you the fish of a lifetime if something goes wrong. If you do a lot of saltwater trolling or fish impoundments with high mineral content, corrosion is an issue you will have to be constantly aware of.

Once you have sorted out the basics of a downrigger look at the other features offered by the many different manufacturers. As with any manufactured product you generally get what you pay for! All the different brands of downriggers will have a range of different features to choose from. Manual downriggers are produced with the crank handle either on the side or top of the downrigger. Personal choice and what you get used to using will always play a part in any buying decisions. The following section looks at the basic components of a downrigger and how different manufacturers have chosen to resolve the design issues involved in the production of this valuable fishing tool.

Line Counters

I consider a line counter to be an essential part of the downrigger. The counter needs to be located near the wheel and easy to read. It provides an easy method of determining depth, though by no means the most accurate. I'll explain this more in detail in the chapter on depth. Most downrigger line counters are driven by a series of pulleys and belts or are gear driven.

Base Plates and Mounts

All downriggers come with a base plate as a means of attaching the downrigger to your boat. The main function of the base plate is to allow removal of the downrigger for storage and maintenance. Think carefully about how you will fix the downrigger to your own boat. Most manufacturers use their own particular system for base plates and are generally not compatible with other brands. If you fish from a small boat that does not have wide side decks you

Downrigging Techniques 79

Scotty Manual Downrigger

may need to use a different type of mount, such as a rail mount.

Most makers of downriggers produce a range of different mounts to accommodate their base plates including deck, side, rail, gimbal, clamp on and swivel bases. Swivel bases make life much simpler when it comes time to set up your downrigger or anytime you are near a jetty or dock. With a fixed base you have to be very careful around other boats, jetties etc because the downrigger arm can be protruding 600 mm or more off the side of the boat. To remedy this you have to remove the downrigger or with a swivel base simply swivel the downrigger so that it is parallel with the side of the boat. Scotty downriggers also have another handy feature in that they have the ability to be tipped up. This feature allows you to change downrigger weights, release clips and set up lines without having to lean out over the side of the boat. You can really appreciate this feature when the water is really choppy or rough.

DOWNRIGGER WIRE

Most downriggers come spooled with between 20 and 80 m of braided stainless steel wire. Small downriggers are generally spooled with 60–100lb breaking strain and large full size downriggers with 130–150 pound. Wire heavier than 150lb is really not necessary and only creates more drag in the water. Your downrigger's wire has the same basic reaction of fishing line cutting through the water. The larger the diameter of the wire the greater the resistance, and the more drag that is created when trolled. This extra drag through the water creates more blowback of the downrigger weight, which introduces more inaccuracy in what the depth counter reads and your real depth

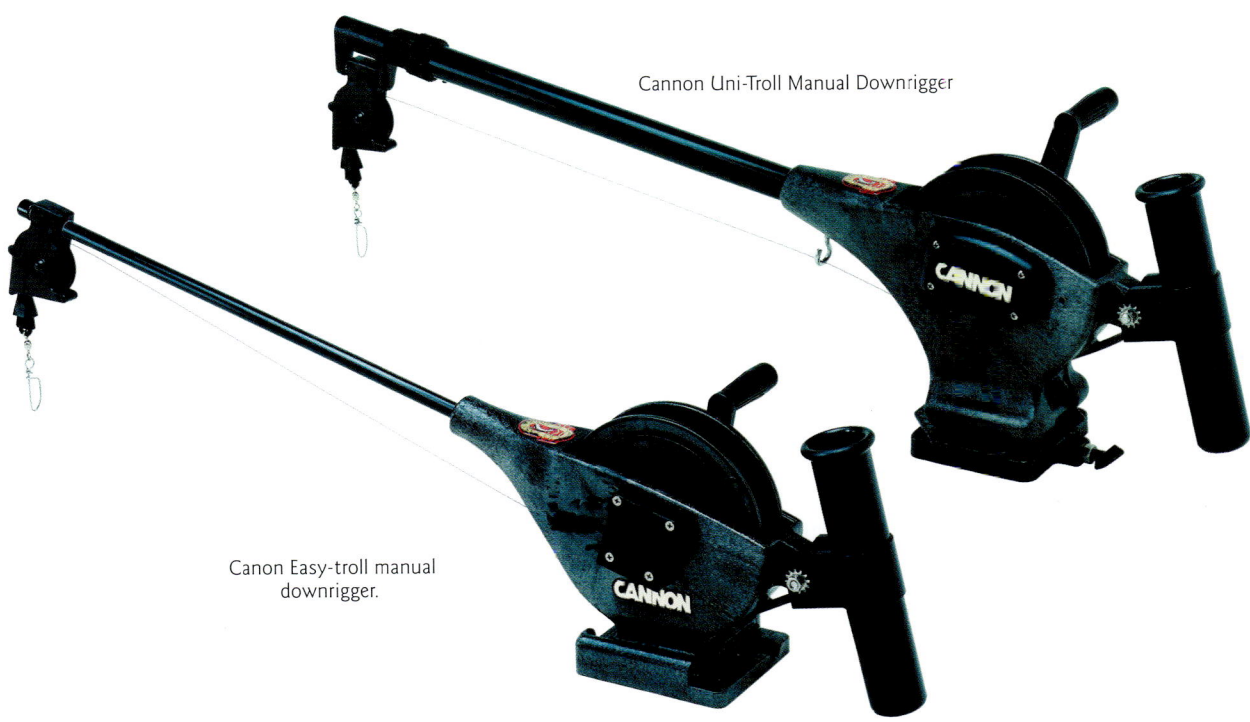

Cannon Uni-Troll Manual Downrigger

Canon Easy-troll manual downrigger.

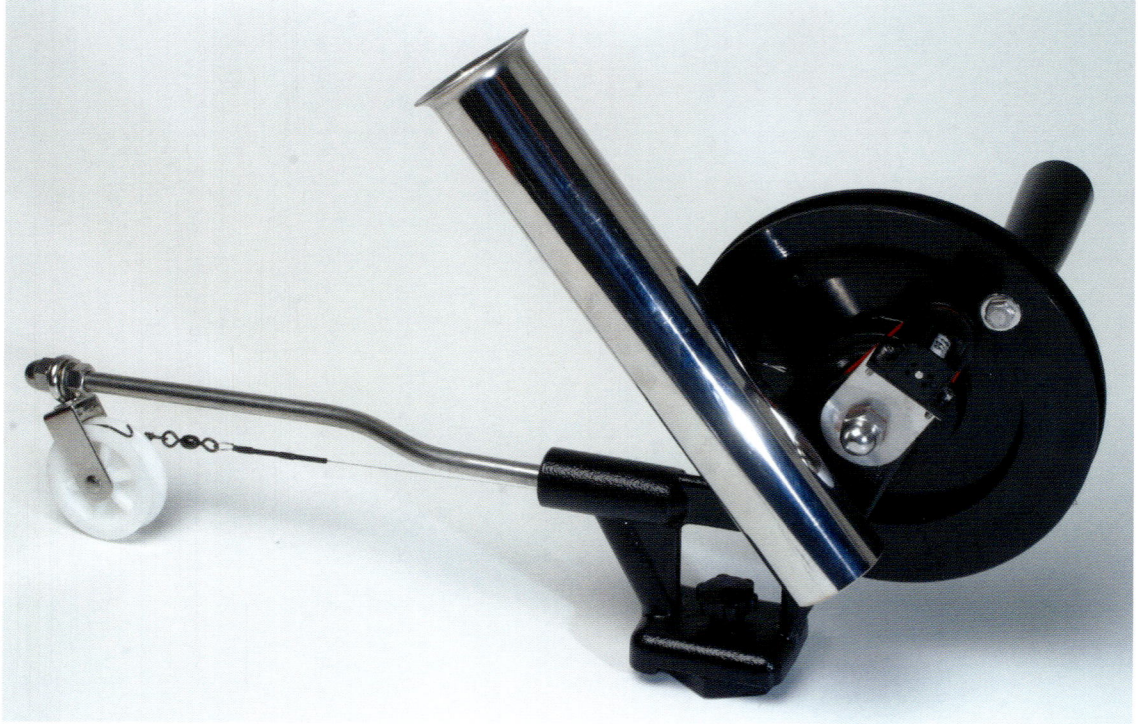

Magnum "Mini" manual downrigger.

The downrigger body is the main support structure to carry the boom or arm, spool for wire and the rest of the downrigger. Most models are constructed from fairly rigid aluminium or moulded plastic or a combination of these two materials.

Boom or Arm

The downrigger boom or arm is generally constructed from aluminium stainless steel or polycarbonate or in some cases a combination of these materials. The arm supports the end pulley wire and the downrigger weight. The amount of leverage exerted on the arm with a 4 or 5 kg weight hanging from it is considerable so the arm must be very strong. Many brands of downriggers offer the option of different length or even telescopic arms. A long arm on the downrigger can be very useful when running multiple downriggers on a large boat, but for most applications an arm length of 600 mm is usually adequate. Keep in mind that as the length of the arm increases so does the leverage on the arm. Downrigger arms of 750–1200 mm are great on big boats, but can pose a real hazard on small boats. I usually carry a pair of heavy leather gloves to try and free the bomb if it hangs up on a tree or rock. Always carry a pair of wire cutters and if you can't free the bomb cut the wire.

> *Safety Note:* No matter what size boat you fish from, never attempt to free a snagged downrigger weight by pulling it off with the boat's power. It is a recipe for disaster, with a very high likelihood of capsizing or sinking your boat.

Electric Downriggers

Electric downriggers make downrigging fast and very efficient. The ease with which you can retrieve the weight, get rigged up and back in the water with your lures is a real bonus for anyone who is serious about downrigging. Most manufacturers produce electric downriggers, with some companies offering several different models. One very useful feature for an electric is an auto stop. This feature allows the angler to retrieve the downrigger weight by activating the retrieve and to continue playing the hooked fish while the weight comes up to the surface or any other predetermined depth and stops automatically. This feature eliminates the potential problem of a fish getting tangled in the downrigger cable when you get the fish to the boat.

Cannon Downriggers make some very sophisticated models that are capable of being programmed to raise and lower the downrigger weight at predetermined cycles of both time and depth. Utilising this feature can be useful if you have fish following the bomb for long periods of time. The Digi-Troll IV downrigger is a very high tech piece of equipment that is loaded with useful features, but it also carries a substantial price tag. Scotty electric downriggers share a common feature with their manual models in that the boom can tip up for easy access to the bomb and release clips.

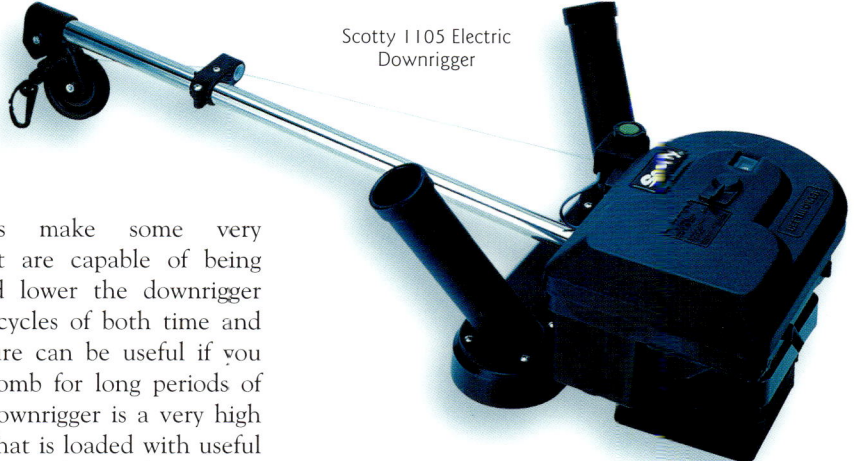

Scotty 1105 Electric Downrigger

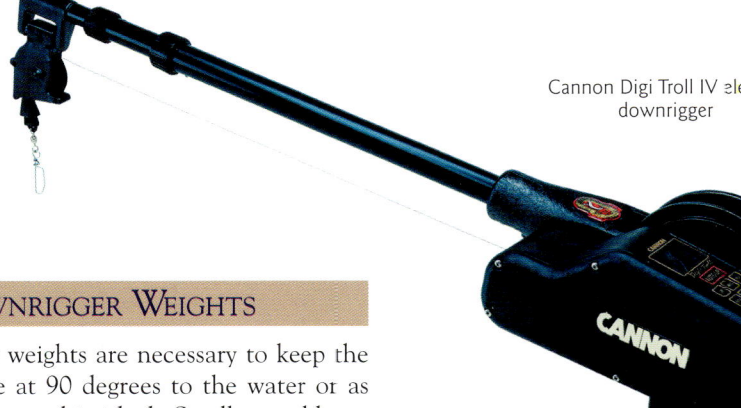

Cannon Digi Troll IV electric downrigger

DOWNRIGGER WEIGHTS

Relatively heavy weights are necessary to keep the downrigger cable at 90 degrees to the water or as close as possible to this ideal. Small portable or clamp-on models used to fish depths up to 8 m often utilise 1.5–2 kg weights and certainly fill a niche. The depth you will troll, your boat's speed and any underwater current you may encounter determine the size of a downrigger weight you will need.

Downrigger weights are often referred to as 'cannonballs' or 'bombs', as the round ball shape was one of the first shapes used as a downrigger weight. Today, weights are made in a range of shapes as well as weights and sizes (see photo of examples). Most shapes of downrigger bombs will do the job; just make sure that the weight tracks true, without swaying from side to side. If you use a cannonball shape without a fin, never run your line release from the weight. If the weight spins it can and will cause a real tangled mess. Stack your release on the cable above the weight.

The photograph above depicts some of the many different shapes available for downrigger weights.

I've used just about every shape of weight that is made and can't really express a preference for any particular shape. I frequently use fish-shaped bombs and have used cannonball and torpedo-shaped bombs for years. I also have used painted fish-shaped bombs, but I still have reservations about whether they actually outperform unpainted weights. At depths below 15 m, I prefer painted bombs, purely to provide as much attraction as possible in the low light levels of these depths.

The issue of environmental pollution is a problem all anglers need to be aware of. Lead is a topic material that can cause pollution in our impoundments, rivers and the oceans of our planet. In some parts of the world its use as sinkers, weights, etc, have been severely curtailed or completely banned. Downrigger weights are not a big turnover item for manufacturers or retailers, but an alternative to lead will happen eventually. In the U.S. and Canada, Scotty Downriggers produce a cast iron weight that they call a Hyperflow weight. Produced in a range of sizes, the Hyperflow is plastic coated to prevent corrosion and is a viable alternative to lead.

Another alternative to lead weights is the Z-Wing downrigger system. The Z-Wing is a lightweight (approx. 325 mm X 175 mm) piece of virtually indestructible polypropylene plastic. This design creates significant downward force that helps overcome drag, allowing a constant depth to be maintained over a wide range of trolling speeds. The Z-Wing has three positions for attaching to the downrigger cable. The centre hole is generally the most efficient and will allow the unit to be trolled at speeds from under 1 km/h to well over 8 km/h.

Downrigger Positioning

Once you've acquired a downrigger or two the decision has to be faced as to where and how you'll mount them on your boat. Setting up a large boat (6 m or larger) is fairly straightforward, but most anglers fishing from small aluminium or fibreglass boats will have a different set of challenges.

In most instances downriggers are generally located on or across the transom or near the transom on the gunwales. Transom mounted downriggers usually extend straight out the back of the boat parallel to the gunwales. Keep in mind that if you run this set-up the downrigger booms must be long enough to clear any trim tabs, extended pods etc.

The Z-Wing System makes an ideal replacement for lead downrigger weights.

Downrigging Techniques

Trolling Boards such as this angler-made version can provide a great deal of versatility when setting up your boat.

Don't get locked into the notion that the only place for a downrigger is the transom. If you have a small boat (3–4 m) that is tiller controlled it may be better to mount your downriggers on the gunwales forward of the transom. With tiller steer boats this makes a lot of sense as the driver is facing forward and can watch the downriggers easily, as well as any passenger. I've had a couple of boats set up like this over the years and the system works well.

The type of mounting you choose for your boat will depend on your boat's gunwales, interior arrangement, freeboard, your budget and personal preference. Pedestal mounts, fixed bases, swivel bases and any other bolted or screwed on mounting should provide years of service. The one drawback with doing this is the drilled holes in the transom or gunwales, not a pretty sight when you remove the bases.

If your boat does not have wide gunwales or side decks you can still mount your downriggers by using a rail mount or rod holder mount. If you use one these types of mounts look at it very carefully.

To mount a downrigger securely may mean altering a commercially made mount or making your own. There aren't many more disheartening feelings than watching a downrigger disappear over the side of the boat because something has broken or come loose. A simple lanyard (I use plastic coated stainless wire) fixed to the downrigger and attached securely to the boat can be an easy remedy and save you the cost of a new downrigger (not to mention valuable fishing time).

An easy method of fixing downriggers to your boat and maintaining some flexibility is to use a trolling board. These boards can be made simply with timber or aluminium and fixed to a rail, pedestal or directly to the gunwales or side decks. This type of mounting board is often a very versatile means for mounting multiple downriggers or accommodating a set up in a small boat. If you use your boat for a range of fishing applications, rod holders and downriggers can often get in the way. A trolling board can give you the flexibility to move all this gear out of the way with a minimum of fuss.

Fish species that suspend or school at depth such as this Australian bass are prime targets for downrigger fishing.

Tackle for Downrigging

Choosing a rod and reel for downrigging will depend on a number of factors including the species you want to target, your choice of spinning or overhead reel, and the decision whether you want one rod to suit everything or a variety for different applications. Downrigging is an excellent means of catching a whole host of freshwater species, but this technique like many others has a few specific requirements.

Rods for Downrigging

The single most important feature to look for in a downrigger rod is the action. To check if a rod has the right action for downrigging you need to bend the rod to see if it will form a parabola (a definite C shape). The safest and most reliable means of doing this is to run line through the guides and get a second person to load the rod by pulling on the line. This little test will not only show you how the rod bends, but also determine if the line touches the rod under load. If this happens it means that the rod does not have enough guides. The reason that this bend is important comes down to 'loading' the rod while downrigging. Loading refers to putting a bend in the rod when it is connected to the downrigger. Water pressure on the line creates a belly in the line. The purpose of the loading of your rod is to remove this 'belly' when a fish strikes and releases the line. This is generally not too big an issue with small fish, but could cost you the fish of a lifetime, if you have too much slack in the line.

Current technology has given us a wealth of information and materials for rod construction, and all this information has added to the confusion that confronts the novice in trying to choose the right rod for this application. Providing your downrigging rod loads up well the choice of length, weight, composition and handles, are all a matter of personal choice. I have a preference for longer rods for downrigging, the longer the rod the more belly (slack line) the rod can pull out of your line when it releases. Longer rods also give you more control

Downrigging rods need to be long enough to pull up any slack line when a fish strikes your lure and have the right action to be able to load enough for this task.

Landlocked Atlantic salmon respond well to trolled lures. Impoundments like Lake Jindabyne in N.S.W. are regularly stocked with these fish.

over a fish during that critical time near the boat. For freshwater use in Australia, I prefer rods in the 7ft–8ft (2.1–2.4 m) range.

The choice of material when choosing a new rod is a fairly complex choice. The new technology has given us choices of materials like graphite, Kevlar and Titanium. The biggest advantage with these materials is their light weight, small diameters, sensitivity and responsiveness. Unless you want one of these rods to do double duty as a casting rod then maybe some of these factors are not going to be crucial in your decision. As nice as these rods are for casting, the trade off for graphite is going to be the price. The other disadvantage with graphite rods is that they often have a tendency to be a bit too stiff for downrigging applications. There are a few manufacturers who are now producing softer rods for this purpose, but they are still expensive.

Most of my rods have cork grips. I still believe that cork gives the most comfortable grip on a rod. The trade off is that rod holders tend to eat into cork grips and in time will wear them out. Some of the tapes that are now made for grips (both fishing rods and tennis racquets) can help protect cork. If you can't live with this then perhaps Hypalon or some synthetic material may be a more practical choice.

Few, if any, tackle distributors or manufacturers bring downrigging rods into Australia. One of the reasons for this in the past has been that demand has been low, but also most of the rods designed for the North American market are just too heavy for the style of fishing we have here in Australia. Most North American downrigging rods are rated for 12–20lb line or 30lb line (6–12 kg). Let's face it, this kind of gear is just too heavy for Australian bass, trout and golden perch. Companies like Daiwa, G. Loomis, Shakespeare, Fenwick, Quantum and Shimano are producing great rods in the 8–17lb (3.5–7 kg) line class due to the demand for rods of this class for walleye fishing in the U.S. and Canada. If you can't find a suitable rod, ask your local tackle shop or consider having a rod purpose-built. Many current graphite and composite fly rod blanks in the 7ft to 8ft (2.1–2.4 m) 6 to 8-weight class make excellent downrigging rods when built as a spinning or overhead rod.

REELS FOR DOWNRIGGING

Your choice of reels for downrigging is again largely a matter of personal preference. You can use almost any type of reel for downrigging, but if you consider the specialist nature of this technique as you would fly fishing, spinning or casting, the requirements for a reel of this type start to narrow down. I have a personal bias for overhead reels for this application purely because they are the simplest, most reliable reels for this technique. A good reel should allow you to free-spool line out as you lower the bomb, without over running the spool. An overhead reel with a bait alarm or clicker will allow you to do this easily. With most spinning reels you either have to flip the bail (generally a short cut to tangles) or back off the drag to allow line to come off the reel. Allowing line to release from a spinning reel in this manner can generate line twist as you lower the bomb. If a fish strikes during this process you lose valuable time (and possibly the fish) while you flip the bail or reset the drag. Spinning reels with a baitrunner system can help overcome these shortcomings, but I still find an overhead easier and more reliable.

Regardless of your preference of reel type

A conventional or overhead reel with a good drag system and preferably with a bait alarm is the first choice for downrigging. Pictured above are a few of the many models available.

The condition that fish are able to attain varies enormously from one body of water to another and is largely dependant on the food source available to the fish. Two of Victoria's southern impoundments, Lake Bullen Merri and Lake Purrumbete, have large minnow populations that enable fish such as this healthy chinook salmon to attain excellent condition in very short periods of time.

some features you need to consider include a strong smooth drag, good line capacity (I consider a capacity of 150–200 m as a minimum for fresh water). There are a wealth of quality rods and reels on the market in both spinning and overhead configuration. Makers such as Daiwa, Abu Garcia, Shimano and newcomers to the Australian scene such as Quantam, Zebco, Okuma and Tica all have excellent products to choose from.

The benchmark for downrigging reels for many years, at least in the North American market has been the Daiwa line counter reels. Like most of Daiwa's reels these are smooth operators, with great line capacity, a bait alarm, line counter, and a good drag system.

The only problem with this reel is that it is just too big for our application here in Australia. Daiwa has now released a new range of line counter reels that are more appropriate for our conditions. The new Accu-Depth range of line counter reels were released in the U.S. market to target the huge walleye fishing scene in that country, and are more of a useable size for our conditions. Companies such as Okuma, Quantum, Shimano, Mitchell, Penn, Shakespeare and Abu-Garcia manufacture good quality line counter reels.

Conventional overhead reels that will also do the job well (some of these reels don't have line counters or clickers) and that are a suitable size include the Abu-Garcia 4600 or 5600C4, Daiwa Millionaire CVZ or CVX, Okuma Convector, Shimano Calcutta and the Iron baitcaster from Quantum. Not all of these reels have a bait alarm or clicker, which means you have to back off the drag to lower your presentation or run the risk of a tangle if you allow them to free spool when lowering the downrigger. If you choose to use a reel without a line counter the Tackle-Tracka is a great way of keeping track of how much line you have out. Australian designed, the Tackle-Tracka is a compact

LCD unit that is fixed to your rod and with your fishing line running through it, allows you to have an accurate measurement of how much line you have out at any time, regardless of your choice of spinning or overhead reel. This device will give you the advantages of a line counter reel, (and a host of other information as well) and give you the option of using whatever type of reel you choose.

If your preference runs to spinning reels for this application, the baitrunner style of reel deserves serious consideration. This type of reel has the advantage of being able to free spool line out as you lower the downrigger, without creating tangles. With conventional spinning reels you need to open the bail to release line (usually resulting in loose loops of line that can cause problems). Another alternative is to back off drag and lower the line working against the reel's drag, which can also contribute to line twist. Baitrunner style spinning reels are produced by a range of companies including Daiwa, Shimano, Penn, Okuma and others. Obviously, you can use a conventional style spinning reel, just be aware that they do have a few drawbacks as mentioned previously.

Using Downriggers

Until the advent of downriggers in the late 1960s in North America most anglers used gear such as lead core line, wire line, paravanes, diving planers or sinkers to get their presentation down deep. These devices have one common drawback in that the weight needed to get the required depth can spoil the fun of fighting a fish. Even worse, increasing the trolling speed or bait/lure depth means that you had to increase the weight or size of the devices, as well as the tackle used and the problem is then compounded.

The best method to attain a given consistent depth is to use a downrigger. In its simplest form, this is just a weight attached to a spool of wire line, and a release clip attached to either or both of them. The clip holds the fishing line (with lure attached) at a certain depth until a fish strikes. It then releases the fishing line from the weight and wire line, allowing you to fight the fish freely.

There are a lot of other reasons why downrigger fishing is better than the previously mentioned techniques for getting to fish that are located deep

If your passion is for catch and release fishing, even fish caught through downrigging need to be handled carefully before being returned to the water.

90 Freshwater Trolling

in the water column. Repeatedly setting a lure at the same depth is easy with the downrigger. This method gives you precise control over the depth of your presentation. If your downrigger has a counter, you just note its reading at the depth you want, and return it back to that reading the next time. If there is no automatic counter, just count the number of revolutions of the spool handle or, even simpler, just read the track on your depth sounder. Most other methods are hit-and-miss and imprecise, particularly when a distance of only a couple of metres can mean the difference between a strike, and your lure or bait being ignored.

With downriggers you can spread a number of lures out over the vertical plane and have a better chance of locating the fish. If a pattern begins to emerge that fish are holding at certain depths, you can concentrate exactly on these hot spots. Vertical hot spots can coincide with thermoclines, temperature breaks, visibility, clarity changes and levels of dissolved oxygen.

Another advantage of using downriggers is that you can avoid problems with floating debris and weeds, especially during autumn. When trolling surface lures the debris or weed catches on your line and then slides down the line to foul the lure. By downrigging just under the rubbish you can avoid the problem, as most weed and debris will catch on the downrigger cable, and therefore not on your fishing line.

Release Clips

A good release clip must be used if you are to be successful when downrigging. For trout and native angling you need a delicate strike mechanism that will release line when there is only a very slight pressure from the line yet the mechanism must also be sufficiently strong to hold the clip closed against the drag of the lure or bait. Some lures require a light release and other lures require a stiff release to set the hooks. Big fish need a different release pressure to small fish. This combination is available with very few types of release clips.

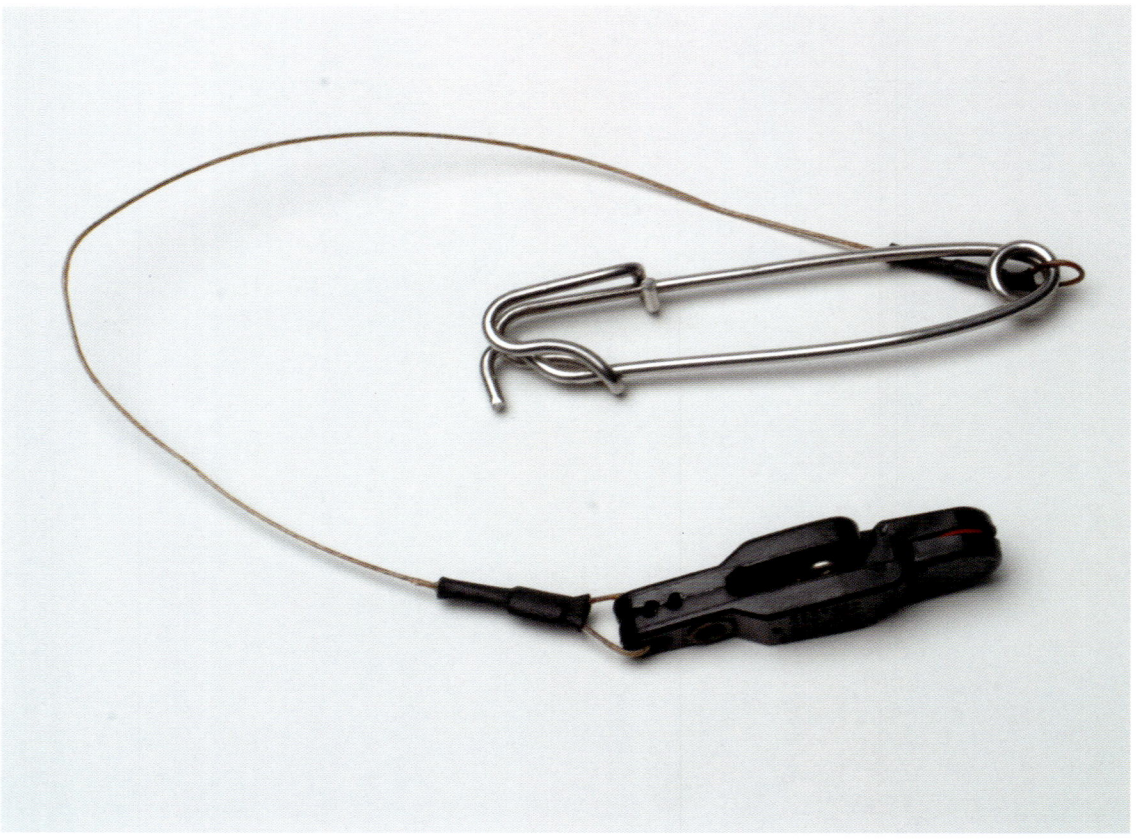

Offshore (Magnum) Release Clip

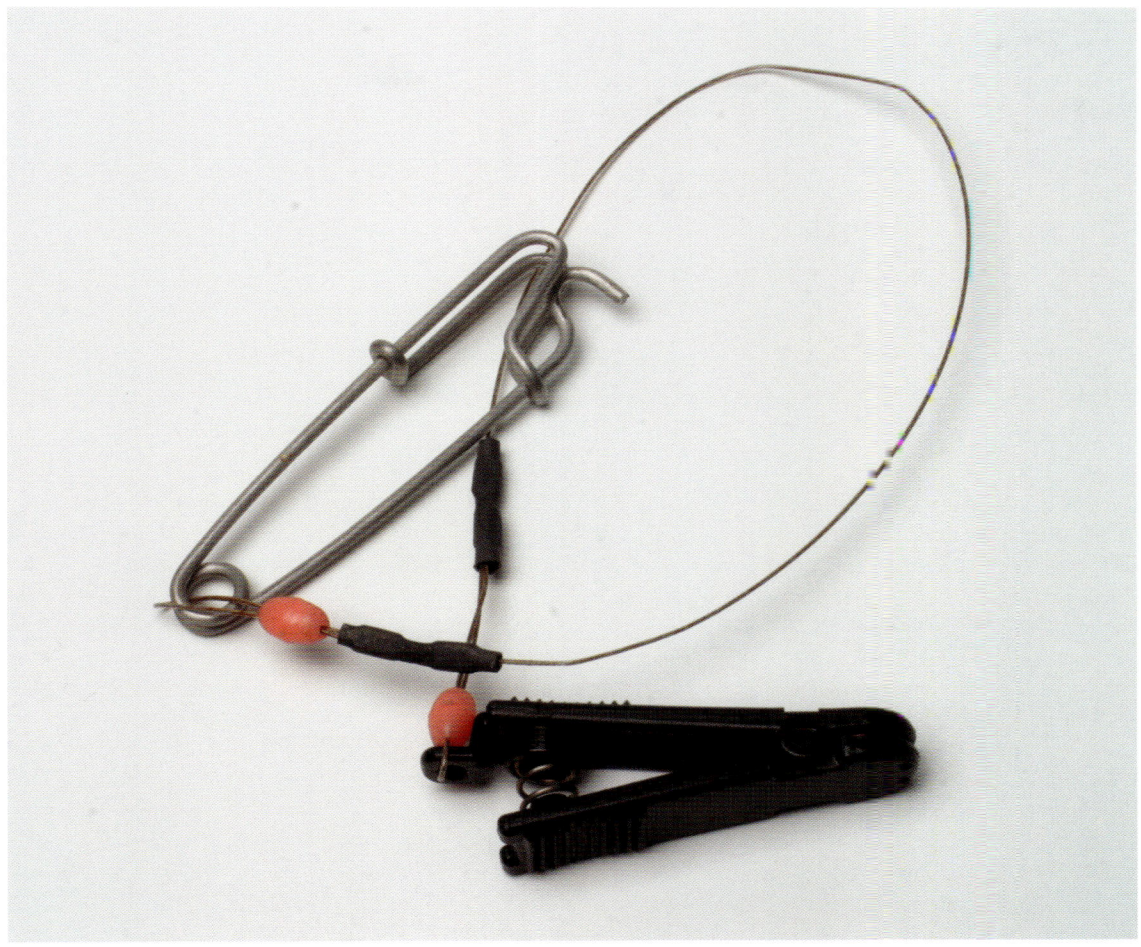

Scotty Snapper Release Clip

'CLOTHES PEG' STYLE RELEASES

Some clips, like the Scotty Mini Snapper Release and Offshore release clips, are basically like a clothes peg in that the line is clamped between two jaws that are usually coated with rubber to try and prevent line damage. These clips are effective but need careful adjustment with lighter line. One problem is that to be consistent with the release pressure, you need to put the line into the jaws exactly the same distance each time. This task is usually difficult to achieve while you are on the water. However many anglers find these release clips the easiest and most convenient to use. Both the Scotty and Offshore clips have been specifically developed for lighter lines between 2 and 4 kilograms.

Ensure you pay close attention to loading the line into these clips and train yourself to be consistent in setting it. As an aid to working out exactly the right position of the jaws for your line class and lures, make a mark on the rubber of the jaws with a marking pen so that you have some reference for the next time you set the clip. On the plus side these clothes peg type clips are quick to set and are uncomplicated.

BUTTON STYLES

These clips are made in Australia by Magnum and are also produced by Mac Jac. This type of clip relies on a plastic button that is threaded onto your line and is then inserted into a wire jaw. When a fish hits, the button is pulled out of the wire jaw and remains on your line. Physically opening or closing the wire jaws to alter the tension adjusts the release tension on button clips. To set these clips hold the

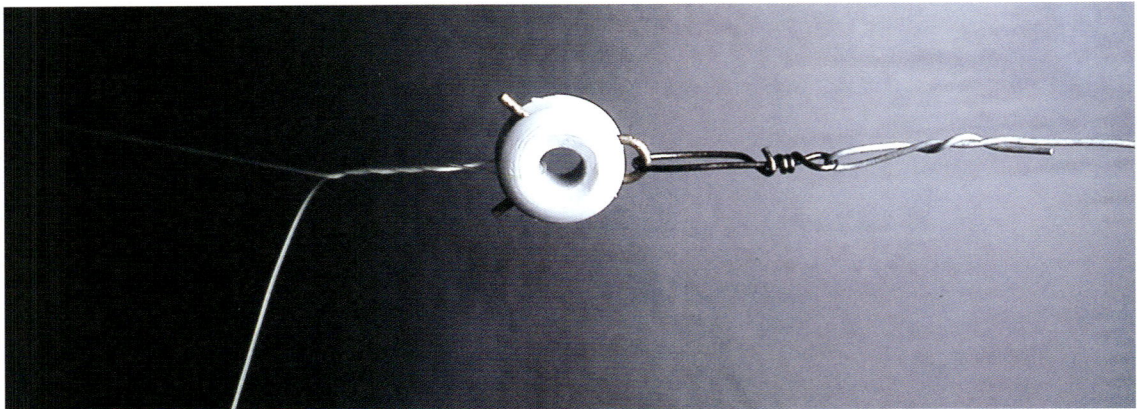
Cannon Release Clip

button in the hand and let the lure trail out the back. When the desired drop back is reached, twist the line five times and snap into the wire jaw. The line has to be twisted so that it will not slip through the ring and this takes time and two hands.

Adjustable Screw and Peg Styles

In my experience, the best peg style release clips are manufactured by Scotty and utilise a completely different principle to other clips. The Scotty Hair-Trigger Release Clip has a peg with a screw adjustment that is very precise and easily adjusted to suit a wide range of release tensions. With very little effort you can adjust the peg to release exactly to suit the lure or bait. A tapered peg that fits into a slotted tube captures the fishing line. It is the resistance between this tube and the peg that dictates how much resistance there will be to the release. This sounds complicated and inefficient but in practice it works extremely well and is very consistent.

Another huge advantage is that the peg is at right angles to the line leading up to the rod. This seems to stop the tension of the line from popping the clip while trolling along, though it releases instantly to a fish. The reason why it works is that the pressure required to release the clip is much

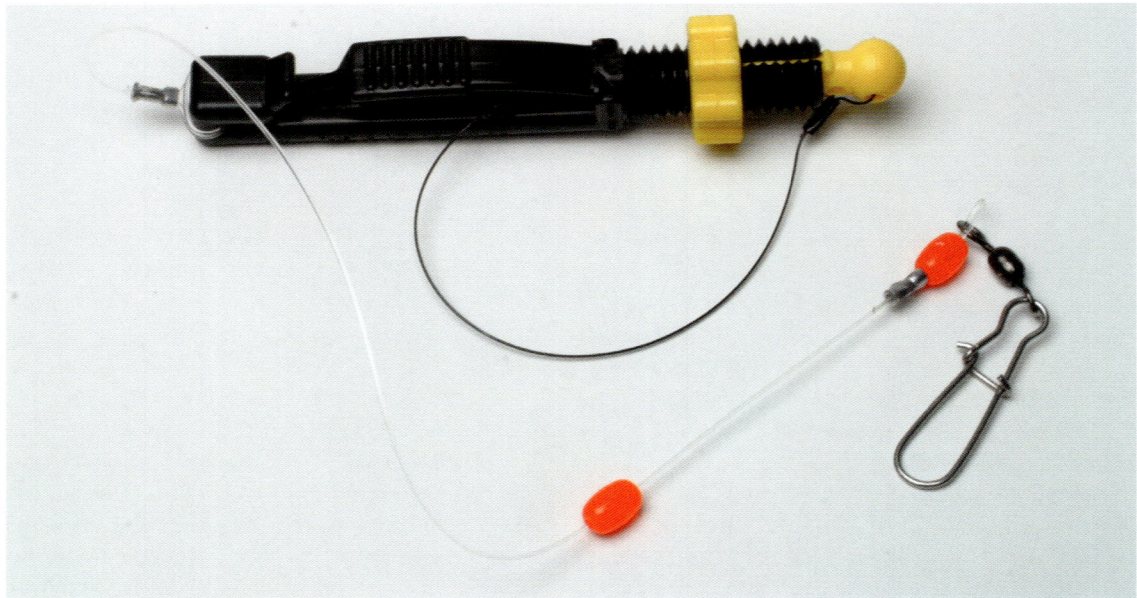
Scotty Hair-Trigger Release Clip

Downrigging Techniques 93

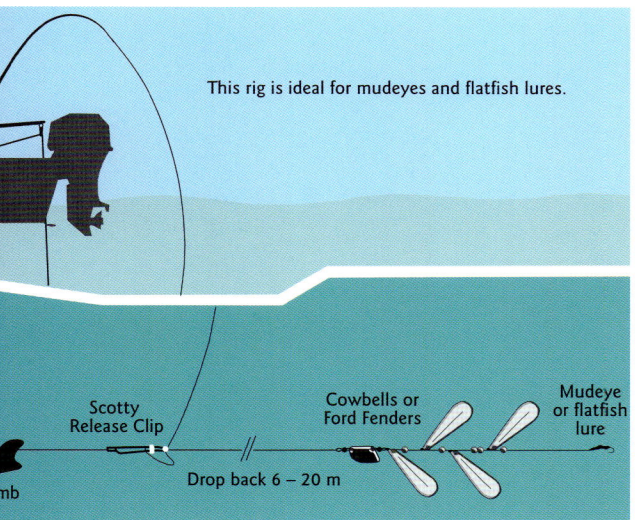

Figure 8.1: Slow trolling attractors with Scotty Release Clip

and reliable and there is little likelihood of error as each setting will release at the same pressure. This reliability and consistency make the Scotty release clip one of the best clips available for Australian conditions.

One small problem with this type of clip is that when the line releases it can sometimes catch on the crimp where the trace to the peg connects to the release clip. This can be avoided two ways. The easiest method is to actually mount the peg of the downrigger clip directly on your line by running your line through the hole in its head where the trace went. This is 100% foolproof but as the peg stays on the line when the clip trips, you can lose the peg if you get snagged. A short piece of Dacron or gelspun line, trimmed tight to the knot, can be substituted and will eliminate this problem on release.

EXTENDING THE RELEASE CLIP

Watching the action on the rod tip while downrigging is just as important as in any other type of trolling. It is always best to have the release clip

less when exerted through the lure tow point than from above, via the line at 90 degrees. You do need some practice to set the Scotty Clip (as you do with all clips) however you will find that it is easy

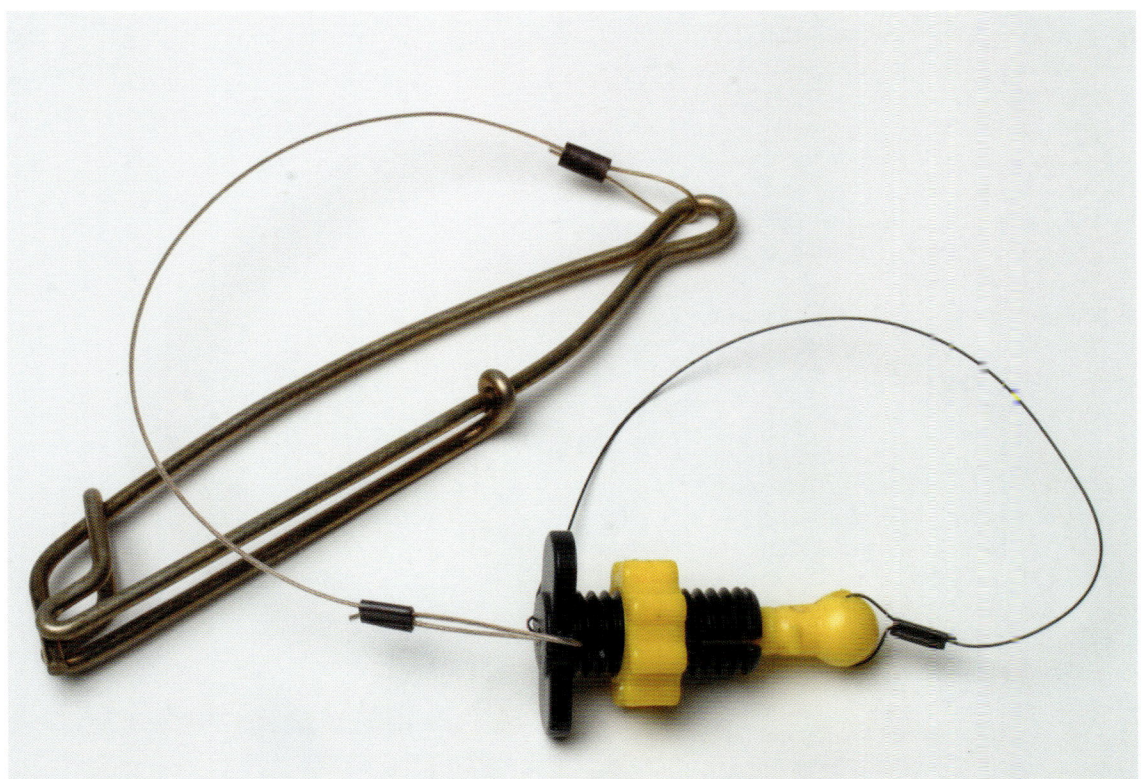

Extended Release Clips such as the one depicted give the angler an indication when a lure has been fouled by weed or a small fish has failed to pull the line from the clip.

on a short wire or Jinkai leader of 30 to 40 cm as it makes an excellent bite indicator and also helps show when you have hooked a fish that has failed to release the line from the clip. By putting tension on the line between the clip and the rod tip, the release clip is pulled up towards the rod tip that is bent over from the rod holder above. When a fish strikes or mouths the bait or lure, this is immediately evident on the rod tip, just like a bite in bait fishing. Detecting this movement will mean the difference between catching fish or not. The vigilant angler will note the tiny movements generated by an interested fish and will take action to arouse it into striking properly. Other options would be to jiggle the rod a bit, change direction or alter your speed. These tiny movements also help to detect bites on sliders when a fish may take the lure and just shake the rod without pulling it out of the clip below.

Sliders

A slider is a trace with a lure attached. It is clipped onto the main line via a snap (see Figure 8.2) and slides down into the water. Water pressure generated by the speed of the boat holds the trace and lure at the middle point of your line, giving you an additional lure placed midway between the release clip below and the surface. When a fish strikes, the release clip below is triggered (or at the very least the rod is given a good shake), and the fish and both lures are brought up.

Stacking and Slider Clips

Two rods or more can be used on the one downrigger but you will need to use two stacker clips to hold the lines of each rod away from the downrigger wire. Stacker clips such as the Offshore (Magnum), or Cannon models, have two clamps, one at each end. One clamps to the wire and the other holds the fishing line. This system allows one to stack lures onto the downrigger wire and run lures at fixed depths (see Figure 8.3).

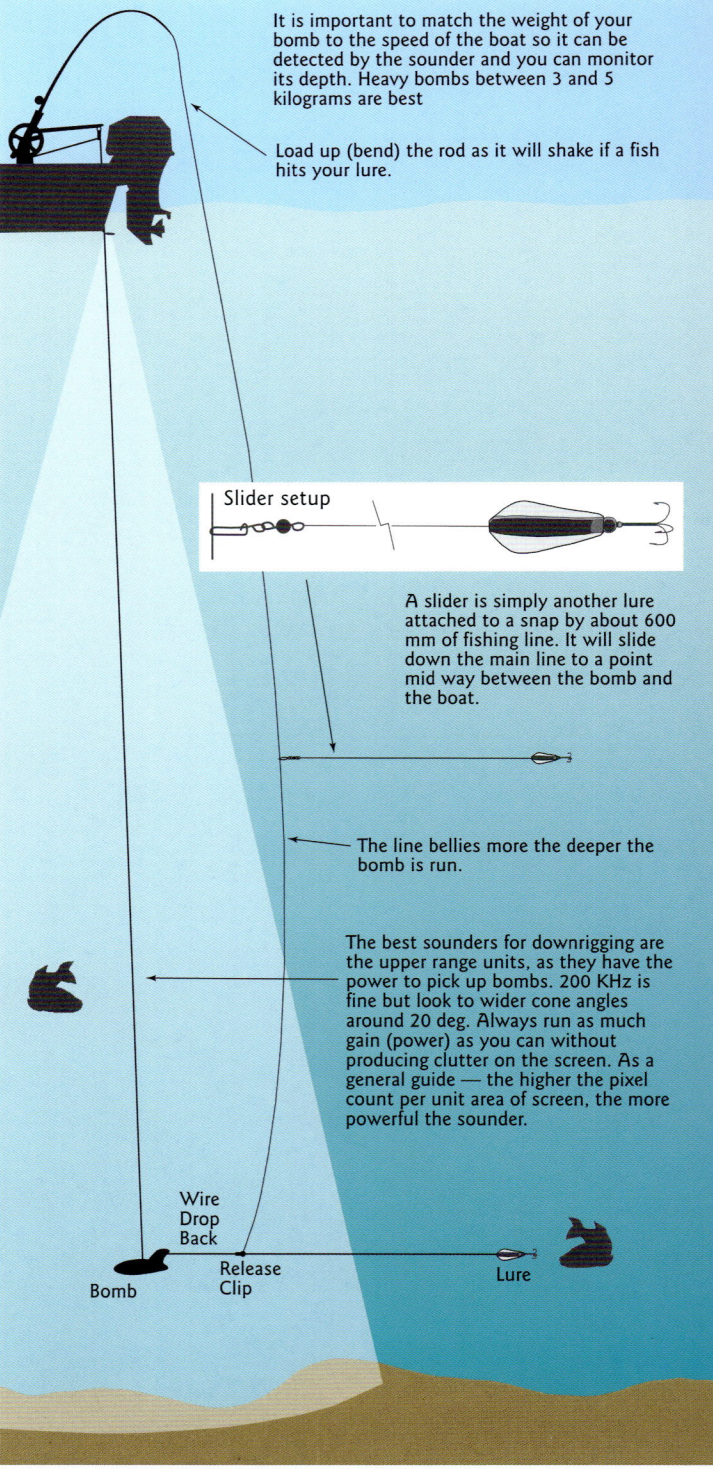

Figure 8.2: Basic Downrigger setup.
Insert: Slider setup.

Downrigging Techniques 95

Scotty Downriggers make small 's' plastic stoppers that can be left on the wire and wound up onto the reel of Scotty Downriggers because they have a spool sufficiently wide to accommodate these clips. On most other makes of downriggers they catch, and cannot be used.

Unless you are fishing clear deep water out of a large craft with forward controls and you are very experienced, I would suggest that stacking could present some serious problems with tangles. Stacking can provide excellent results when you find a school of fish or fish are located near a thermocline or structure. This technique can be very useful in the right circumstances and with a little practice can be easily mastered.

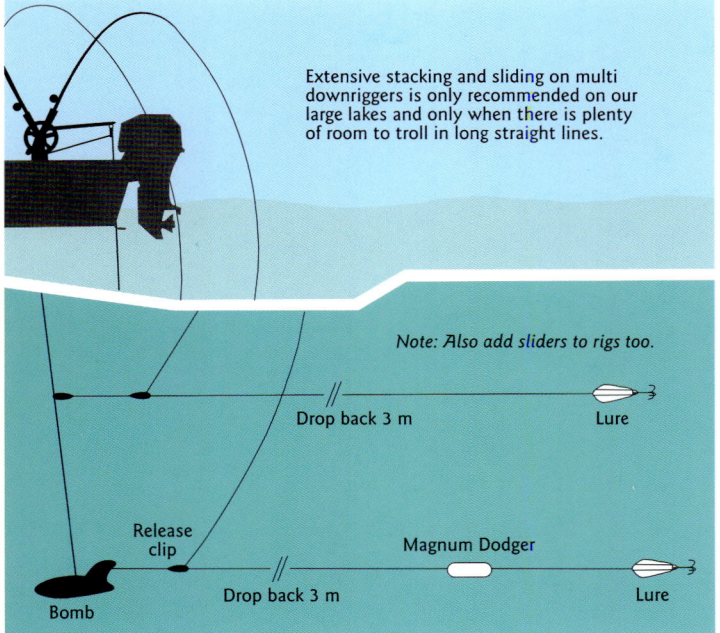

Figure 8.3: Stacker setup

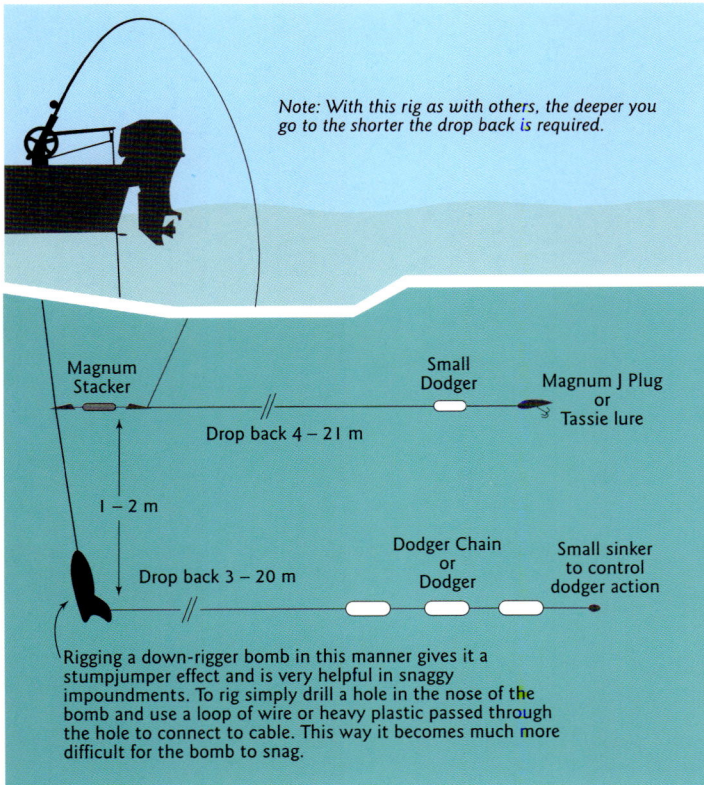

Figure 8.4: Magnum Stacker

Drop Back

Perhaps the most common question anglers ask regarding downrigging is how much line to let out before setting the line in the clip. This depends on a number of factors: what style of lure you are using; the time of day and quality of the light; how shy the fish are; and how deep you are setting the lure or bait. As a basic guide, I like about 5 to 30 m of drop back from the bomb to the lures when fishing depths from 2 to 5 metres. From 5 to 10 m trolling depth it should be shortened to 8 to 15 m and from 10 to 20 m deep about 3 to 8 m back. This is only an approximation; fishing conditions will dictate how long your dropback will need to be. I have had many days when a dropback of 1–1.5 m has caught a lot of fish, even at depths of less than 5 metres.

When your targeted species is easily spooked by water conditions or fishing pressure you need the lures further back, yet there are times when the fish will take the lures very close to the bombs. I am sure that they are

96 *Freshwater Trolling*

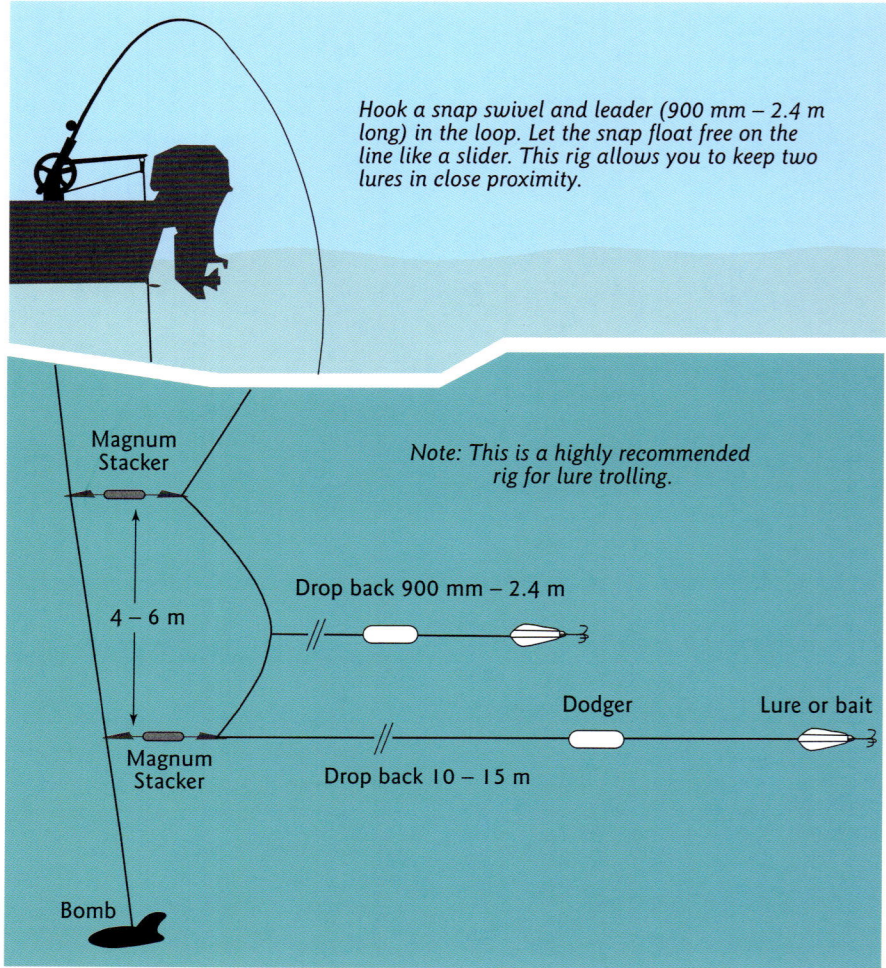

Hook a snap swivel and leader (900 mm – 2.4 m long) in the loop. Let the snap float free on the line like a slider. This rig allows you to keep two lures in close proximity.

Magnum Stacker

4 – 6 m

Note: This is a highly recommended rig for lure trolling.

Drop back 900 mm – 2.4 m

Dodger — Lure or bait

Magnum Stacker

Drop back 10 – 15 m

Bomb

not really worried by the bomb, or the motor or the boat as long as there are no undue sudden noises. Be aware too that fish (including both natives and trout) will investigate the bomb and you will often see them doing this on the sounder.

Cable Hum (Hum of the Downrigger Wire)

As the bombs are dragged through the water at 6 m or more in depth and especially at the faster trolling speeds the wire can vibrate and make a real humming sound. Some anglers feel that it will drive fish away but while the hum may be annoying, there is no evidence that it scares the fish. The wire actually hums all the time but sometimes at a frequency more noticeable to the angler. If the noise

Figure 8.5 ABOVE: Magnum Stacker

RIGHT: In Australia downrigging is often associated with fishing for trout and salmon, but in reality it can be effective for almost any species. If it swims and will take lures or a bait you can catch it with a downrigger!

bothers you, it can be minimised by mounting the downrigger on a heavy sheet of rubber to isolate it from the hull. This works well on aluminium boats. You can also tie a piece of inner tube rubber from the wire to the boat to absorb the vibrations but don't forget to remove the rubber before you start to wind up the downrigger.

Cable Blowback

Cable blowback is the phenomenon that occurs as water pressure pushes against the downrigger cable and bomb while you are trolling. Experienced trollers can tell by the angle that the cable enters the water as to how much this occurs. Blowback can affect how deep your lure or bait is running at a particular speed, as the further the bomb blows back the less accurate your depth counter will be, and the more difficult it will be to predict at exactly what depth your lures are running.

There are instances when you can actually take advantage of this phenomenon. Normally

Like any other technique in fishing, choosing the right gear for downrigging can help ensure your chance of success.

to overcome the problem of blowback you can decrease your speed or increase the weight of your downrigger bomb to stop the blowback. To take advantage of this fact use a lighter weight (1.5–2 kg) bomb and vary your speed, including putting the motor in and out of gear. This technique will allow your trailing lure or bait to flutter down as the weight angles back into the transom of your boat. Very short drop backs of 1–3 m seem to be the most effective with this technique, most likely because the shorter drop back more easily transfers the movement to your trailing lure.

Positive Charging of Downrigger Systems

Some boats catch more fish when downrigging than others and some seem never to get good results. I used to put this down to the experience of the anglers, the equipment used or just good/bad luck.

Overseas research has shown that some species of fish are extremely sensitive to weak electrical fields in the water. Trout and salmon (as well as sharks in salt water) are species that are very sensitive to negatively charged electrical fields.

In North America most commercial skippers using downriggers to troll for salmon have discovered that they need to use marine electronic circuitry installed in their craft to give their downrigger cables a slight positive charge. They even use a voltmeter connected to measure the polarity of the electrical current between the boat and the downrigger cable. Aluminium boats exhibit these voltage gaps more so than fibreglass boats.

Here in Australia, much of our water is fairly mineralised and a situation can arise where the ions in the water produce an electrical current between the dissimilar metals of the boat, motor and downrigger bomb. Trout especially detect these weak electrical fields. If the field produced on your boat has a positive charge around the bomb and wire then you're in luck as it has been found that trout are attracted by a positive charge of around 0.5 or 0.6 volts. If your boat has a negative reading and trout are repelled by a negative charge, then you might consider doing something about it.

There is an easy way to test for negative or positive charges. Simply lower the bomb about 3 m into the water, connect a small voltmeter negative lead to the boat engine and the other to the downrigger wire. If the downrigger wire is positive, then 'you're cooking with gas' and the fish will not be repelled.

You can alter the electrical reading to positive by making sure the zinc anode on your motor is regularly replaced and always in good condition. The sacrificial zinc anode can also become coated with a salt when outboards are kept out of the water on a trailer. This coating will diminish the anode's function so keep it clean for the best results.

Other helpful tips are to use a plastic snap to connect the bomb to the stainless steel wire as this helps insulate the current. Scotty and Cannon both sell their downriggers with a plastic connection clip. I always thought this was just to save money but now I realise that they had this electrical problem in mind when designing their product. Also use bombs made of pure lead, not alloys of zinc, magnesium etc., which can increase the electrical current. You should use a downrigger with a plastic spool, not aluminium or similar metals, as you are trying to insulate the downrigger, wire and bomb from the metal ground of the boat.

ELECTRONIC FISH FINDERS

Modern electronic fish finders allow you to pinpoint the location and depth of the fish and the structure surrounding them; the downrigger then allows you to present your lure or bait to the exact depth of the fish. A good quality sounder will also show the exact position of the bomb so you can watch for snags, submerged trees, or even individual fish and lower or raise the bomb and your lure accordingly.

This ability to be in contact with and respond to what is happening under the water is the secret to downrigger success.

The best makes of fishfinders are also sensitive enough to distinguish a thermocline (a depth where there is a marked variation in the water temperature). Fish often concentrate along these thermoclines as they seek water of a more comfortable temperature and also because baitfish and plankton are found there.

Unfortunately not all depth sounders seem to be capable of such accuracy. For instance most of the 3D models extensively advertised will not show

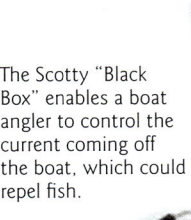

The Scotty "Black Box" enables a boat angler to control the current coming off the boat, which could repel fish.

the bomb under the boat as these units are designed for a different application. I recommend you to go for the most powerful unit in a 2D that you can afford. A depth sounder that will not show a 10lb lead weight under the boat must be missing a lot of fish as well!

To get the best readings from your depth sounder, mount your transducer on the same side of the boat as your downrigger and angle your transducer slightly backwards to put the bomb into the beam angle. Depth sounders with 18–22 degree transducers usually have no problems in showing the bomb but those with narrower cones of 5 degrees need to be adjusted. I have used and tested most of the Lowrance and Eagle models and both brands show the bombs on both sides of the boat without any problem at normal depths and speed. This is helped by the fact that I have my downriggers mounted forward a little in the boat which means that the bombs are directly beneath the transducers while in operation—an ideal set up that helps a lot. Personally I like to see the position of the bomb on the sounder at all times.

LURES FOR DOWNRIGGING

Nearly any lure can be used on a downrigger, but some types work better than others. As the downrigger bomb does all the work in taking your lure to the selected depth, you can use the lightest lures with sensitive actions, such as flutter spoons, and the floating range of minnow lures. Of course one of the most effective lures for trout used off a downrigger is still the range of popular Tasmanian lures such as Wigston's Tasmanian Devils or Lofty's Cobras.

When using a downrigger you don't really need sinking or deep diving lures, use the same style and patterns but in a floating version. Floating divers or shallow running lures are generally lighter and more sensitive, but they also have the big advantage of floating and staying clear of snags while you retrieve lines and get things organised if you have to stop the boat. It is also a lot easier to predict an accurate running depth for your lures. There are of course exceptions to this rule! Targeting fish that are on or near the bottom is always a difficult proposition. This scenario is usually a recipe for losing tackle including downrigger bombs, line releases, lures, and even downrigger wire. To avoid the hassles of losing all this gear you can run a deep diving lure off

Advances in technology mean that modern fish finders have a range of useful features and are very user friendly. The new generation of sounders from Lowrance are tough, lightweight, waterproof and very affordable.

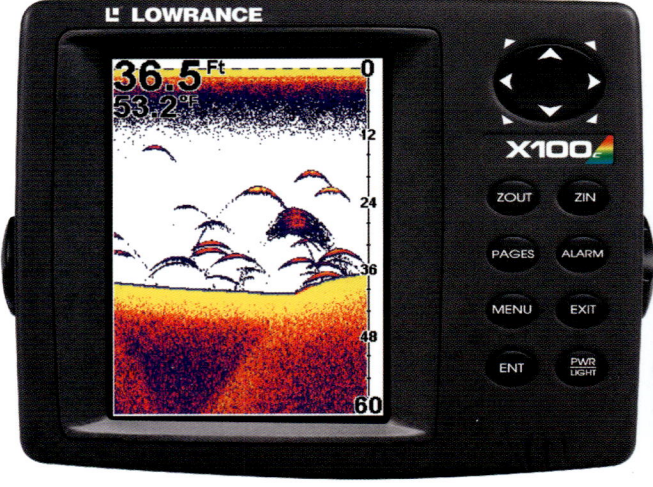

Transducer mounting

Downrigging Techniques   101

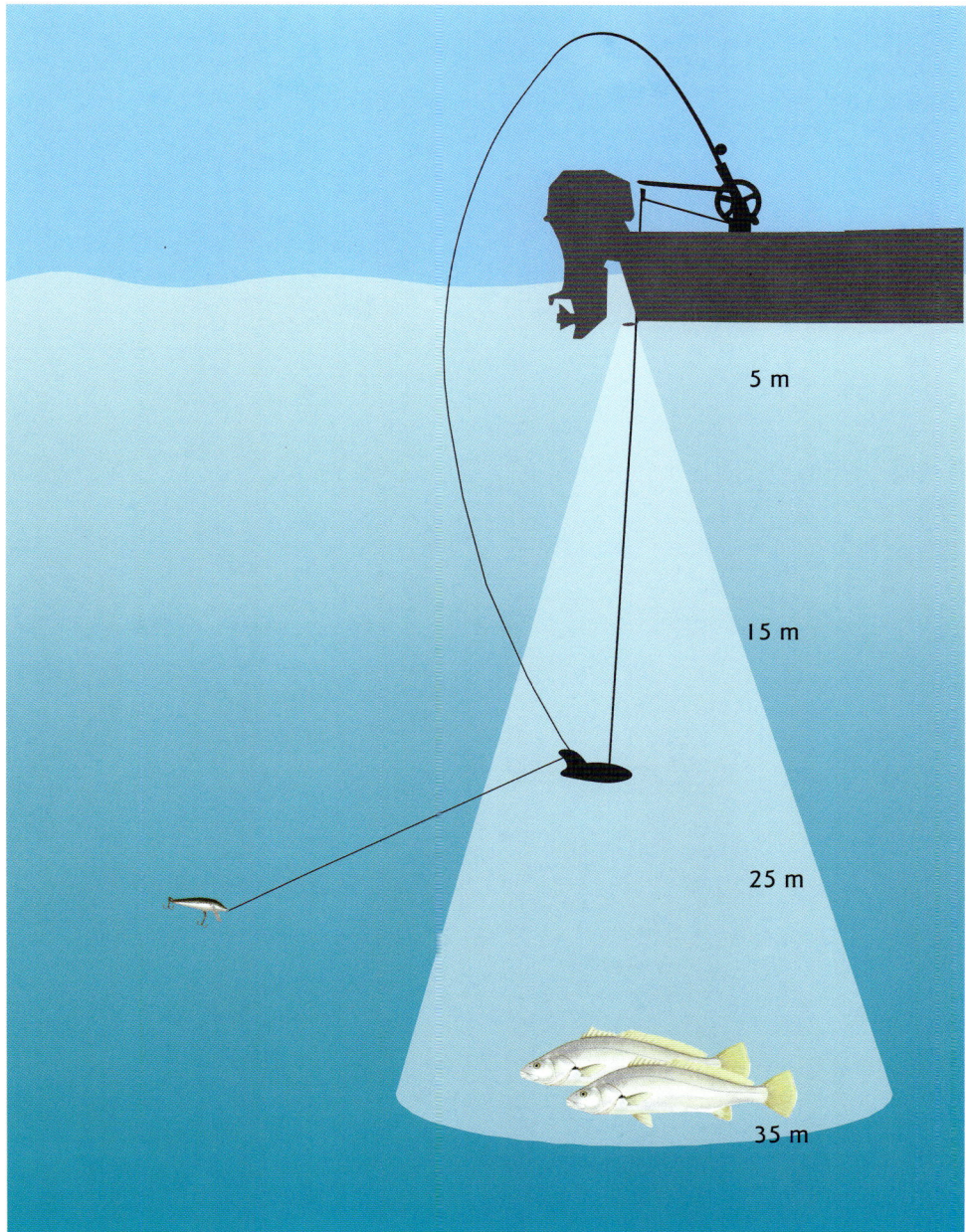

Figure 8.6: Rigging deep diving lures on the downrigger. If you choose to run a deep diving lure off the downrigger keep in mind that the depth the lure will troll is determined by how far the lure is run behind the downrigger weight.

the downrigger. Choose a lure that you are confident will run at a consistent depth behind the downrigger. As an example, if you choose a lure that dives to 4 m and you are fishing in an area with a fairly consistent bottom shape you can set your downrigger to run 5 m above the bottom. This will give you a metre of clearance between the bottom and your lure. You will still have to pay close attention to your sounder and watch what happens to the contours and depth. You may still snag your lure, but it is cheaper and less time consuming to replace a lure than all the other possibilities.

All lures have their place in downrigging and it is more a case of spending time to think about what

Downrigging Techniques

your lures are going to do under water, and trolling at the right speed. Downriggers give you a versatility that was not there when you had to rely on the designed characteristics of the lure alone to reach the desired fishing depths. As a general rule, you can run lures a bit quicker off a downrigger than off a flatline as the water pressure is consistent all the way around it. This is especially true for large hard body minnow style lures. These larger lures can easily handle speeds much greater than you might employ for smaller lures. But be careful of your spoons and Tasmanian Devils as it is easy to run them too fast and cause them to do full 360 degree loops, resulting in twisted lines and tangles. If your lines become twisted while trolling, you can bet you are tolling too fast for the lures. Thin bladed flutter spoons are one of the best choices if you need to troll very quickly, as they can handle very fast speeds without spinning out of control.

Clearing Lines

When a fish is hooked up while you're downrigging with a number of rods out, keep the boat moving forward. Unless you're lucky enough to strike a patch of consistently big fish, don't stop or put the boat into neutral, especially if you're trolling into the wind, as it will result in a memorable tangle! Keeping the boat moving forward will keep everything in order and clear, and in most cases you can actually fight the fish up to the side of the boat and land it without interfering with the other lines. The vibration and flash of the fish being fought can also trigger a strike on one of the other lures at the same time. I usually wind up the downrigger on the side that has had the strike so that this area is clear to land the fish.

It can be a real test of your skill to land a fish with other lines still in the water.

This great brown trout is testament to the fact that downrigging can be productive. This fabulous fish was landed from downrigging in Lake Jindabyne N.S.W.

If the fish is obviously big, then the other rods and the downriggers can be cleared or retrieved and the angler playing the fish then has the opportunity to play the fish without the worry of tangling other lines.

Jigging (raising and lowering) Downrigger Weights

This technique involves raising and lowering the downrigger weight to target fish that you observe on your sonar. Often fish will follow your trolled lure for long periods of time. If you raise and lower the downrigger bomb and your trailing lure this can sometimes be the trigger to get a fish to strike. If you own a manual downrigger this may prove to be quite an effort, but the result can be worth it!

Useful Tips for using Downriggers

- Before setting out your lures always check to see that each lure is working properly.
- Make sure that the spread of lures that you choose to run are compatible and will troll effectively at your chosen speed.
- Always watch your rod tips: they can be the first indication that you have hooked a small fish that hasn't tripped the release or that your lure is fouled with weed or debris.
- Avoid trolling in a straight line. Changes in speed or direction can help get the best from your presentations.

Tactics that will help catch fish

- Troll slowly; most big fish won't expend any more energy than is absolutely necessary to chase a meal. Rainbows will go for a lure or bait at a moderate to fast speed, but browns seem to prefer a slower presentation.
- Don't troll in a straight line! Trolling slowly is the key to success but always vary your speed and direction. Trolling in 's' curves will always make your lure or bait a more consistent producer. Changes in speed and direction produce different vibrations and action from your lures and make them more productive.
- A good quality depth sounder is a must for this type of fishing. The latest range of sounders from Lowrance is hard to go past with the X100C, LCMX 15 and 19 having a battery back up to retain a memory. Sounders enable you to locate structure, drop offs, bottom contours and all importantly, fish.

THE DIVE BOMBING TECHNIQUE

Varying your trolling technique can often give that break in the pattern that can produce results. In this case the change causes the lure to appear like one of the minnows affected by the bubbles being emitted by the aerators, easy prey for the predating trout and salmon.

CORRECT WAY

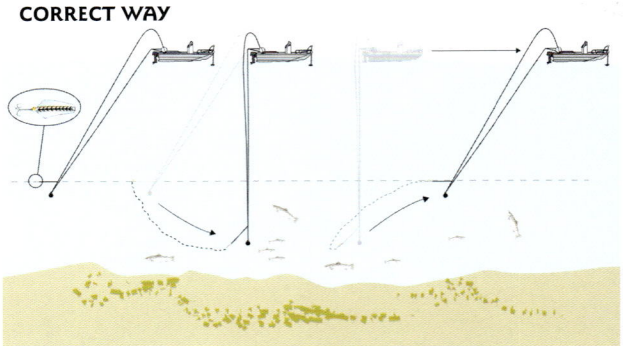

The boat while underway trolling.
The lighter, more water-resistant bomb swings well behind the boat and is actually running 25 percent shallower than the length of cable attached to the bomb. The depth at which this bomb is running is difficult to track on the sounder.

The boat is stalled while trolling to vary the action.
The bomb has a downward and forward action. The lure can flutter downward about 25 percent of the length of bomb cable that is out. The action of the lure is controlled by the short dropback from bomb to lure: trolling very slowly, downward and forward, imitating a wounded or fleeing baitfish. The boat is stalled while trolling to vary the action.

The motor is placed in gear and the boat restarts.
The bomb is held back by the water resistance and swings back and up as it speeds up. The lure is slowly pulled upward and along like a baitfish fleeing the depths.

INCORRECT WAY

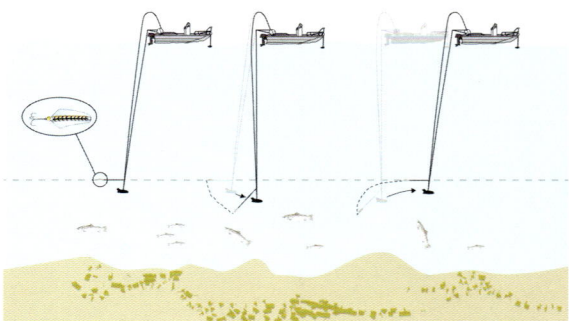

The boat while underway trolling.
The heavier, more streamlined bomb travels almost directly under the boat. It's most likely that this bomb can be tracked by the boat's depth sounder so it is much easier to verify that it is running at the depth of cable that you have out.

The boat is stalled while trolling to vary the action.
The bomb only has a short distance to swing forward and down. The lure can only drop down the length of the dropback under its buoyancy, giving it little or no action.

The motor is placed in gear and the boat restarts.
The bomb quickly reaches its trolling line and action with the lure being pulled upward directly and firmly, about the length of the lure dropback.